N A

I A N A

The Neuroptera of Malesia

FAUNA MALESIANA HANDBOOKS

Series editor A.J. de Winter
Project editor G.O. Keijl

Fauna Malesiana Foundation
c/o National Museum of Natural History Naturalis
P.O. Box 9517
2300 RA Leiden
The Netherlands

The Fauna Malesiana Foundation received financial support from the Royal Netherlands Academy of Arts and Sciences, and the National Museum of Natural History Naturalis, Leiden.

FAUNA MALESIANA HANDBOOK 4

The Neuroptera of Malesia

T.R. New

BRILL
LEIDEN · BOSTON

2003

Library of Congress Cataloging-in-Publication Data

New, T.R.
The Neuroptera of Malesia / by T.R. New.
—(Fauna Malesiana, ISSN 1388-3895 : v. 4)
Includes bibliographical references and index.
ISBN 9004124977 (cloth : alk. paper)
1. Neuroptera—Malesia—Classification. 2. Neuroptera—Malesia
—Identification. I. Title. II. Series.

Further details available from the publisher.

Die Deutsche Bibliothek - cip-Einheitsaufnahme

New, T.R.:
The Neuroptera of Malesia / T.R. New. -
Leiden ; Boston: Brill, 2003
(Fauna Malesiana, Volume 4)
ISBN 90-04-12497-7

This book is printed on acid-free paper.

ISSN 1388-3895
ISBN 90 04 12497 7

Cover and interior design by John Stelck (Biskaje), Leiden

Printed in The Netherlands.

Contents

	Preface	VII
	Acknowledgements	VIII
1	Introduction	1
2	The Neuroptera	5
	Morphology	6
	Biological outline	8
	Higher classification	10
	Key to the families	14
	Collecting and preparation	20
	Economic importance of Neuroptera	21
3	Biogeography	23
	Faunal analysis	29
4	The families of Neuroptera	35
	Rapismatidae	36
	Coniopterygidae	39
	Sisyridae	50
	Osmylidae	53
	Mantispidae	59
	Berothidae	68
	Dilaridae	72
	Hemerobiidae	75
	Chrysopidae	86
	Nymphidae	107
	Ascalaphidae	111
	Myrmeleontidae	122
5	Checklist of Neuroptera from Malesia	145
	Rapismatidae	147
	Coniopterygidae	147
	Sisyridae	151
	Osmylidae	151
	Mantispidae	153

Berothidae 158
Dilaridae 159
Hemerobiidae 159
Chrysopidae 163
Nymphidae 170
Ascalaphidae 171
Myrmeleontidae 174

References 181

Glossary 189

Index 197

Preface

This handbook is a tool to help understand the Neuroptera, lacewings and their allies, of a complex geographical region, Malesia. It is a tentative summary of available information, in which I have attempted to review the recorded taxa whilst pointing out the shortcomings of our knowledge, and to provide keys to help recognition to generic level for all families. Inevitably, some taxa can be included only with considerable uncertainty, and I have noted (for example in the antlions, Myrmeleontidae) the kinds of problems that occur in diagnosing species described very briefly, for which specimens are not available readily for detailed examination, and which have not been reported for many decades. However, by indicating the limitations of information available, and augmenting the text and illustrations with a reasonably full bibliography and checklist of regional species records, I hope that entomologists in Malesia may be encouraged to take up the study of these fascinating and economically important insects. In the introduction I note that one reason that our knowledge is so fragmentary has been the lack of students resident in the region; if this book helps to redress this by fostering local interest and expertise, it will have proved its worth.

T.R. New
Melbourne, September 2000

Acknowledgements

I appreciate greatly the support of the Fauna Malesiana Foundation for the preparation of this handbook, together with the continued encouragement of Dr Ton de Winter as it has developed. I am very grateful for the perceptive and thorough editorial comments of Mr Guido Keijl.

Works such as this can be written only after study of material in collections of taxonomic institutions and with advice from other specialists, and I am grateful for the help, friendship and hospitality received over many years of looking at Neuroptera. In particular, the unrivalled worldwide collections of the Natural History Museum, London, and the magnificent library facilities of the Department of Entomology, have been indispensable sources of information, and I thank Dr Peter Barnard, Mr Stephen Brooks and Ms Julie Harvey for their courtesies and help during my visits to London over many years. Many neuropterists have helped, directly or unwittingly, in this taxonomic synthesis, through responding to queries, sending me copies of their papers and general encouragement; they include the late Dr Phillip A. Adams (Fullerton, California), Univ. Prof. Dr Horst Aspöck and Dr Ulrike Aspöck (Vienna), Dr Oliver S. Flint Jr. (Washington), Dr Herbert Hölzel (Eppersdorf), Dr Mervyn W. Mansell (Pretoria), Prof. Martin Meinander (Helsinki), Prof. Victor J. Monserrat (Madrid), Dr John D. Oswald (Texas), Dr Norman D. Penny (California), Dr György Sziráki (Budapest), and the late Dr Bo Tjeder (Lund).

For permission to reproduce figures from my published papers, detailed individually in figure legends, I thank CSIRO Publishing, Collingwood, Victoria; The Royal Society of Tasmania, Hobart; the Australian Entomological Society, Brisbane, Queensland; and the Finnish Zoological and Botanical Publishing Board, Helsinki.

chapter **1**

Introduction

This handbook is an introductory guide to the Neuroptera of Malesia. The Neuroptera (or Planipennia), the lacewings and their allies, is one of three orders of primitive holometabolous insects conventionally grouped in the superorder Neuropteroidea (or Neuropterida), together with the Megaloptera and Raphidioptera. Although the Neuroptera are not particularly well known, considerable appraisal of the Malesian fauna has occurred in recent years, and this handbook is timely as the first attempt to summarise comprehensively much of the highly scattered information on these insects from the region, and to provide a source for information on their systematics, evolution and distribution.

Unfortunately, and despite the undoubted importance of some taxa as manipulable predators of crop pests, only fragmentary, sporadic biological information is available. The detailed biology of most taxa is entirely unknown. The bulk of literature comprises taxonomic descriptions, with many of the species described rather briefly and known from somewhat abbreviated characterisation and from single localities. In many accounts, related but different species have been referred uncritically to an 'available name' without perceptive examination, and clarifying the extent of the fauna has necessitated reappraisal of type material (where it has been possible to trace this), and careful evaluation of information in original descriptions and diagnoses. Recent studies have also revealed the presence of many undescribed species. With very few exceptions, Malesian Neuroptera have not been subject to comprehensive revisions in any broader context; many isolated taxonomic descriptions of species from the region make only scant comparative reference to other taxa, even from the same country or countries. These comments apply also to many other insect groups.

Nevertheless, a reasonable perspective of the spectrum of taxa present is available, and this allows some evaluation of the region's characteristics and peculiarities, as a template for facilitating future investigation and in enabling workers in Malesia to recognise and identify many of the Neuroptera they encounter and, importantly, to recognise novelties in the region. Many of the species are known only from the region, so that apparent endemism is high. However, as many have been collected on only few occasions, and thus appear to be rare, it is unclear whether this impression is genuine or simply a result of undercollecting. It is likely that many of the species are far more widespread than at present documented, perhaps especially in the contiguous regions of West Malaysia, Thailand, Myanmar, and Cambodia, Laos and Vietnam. Additional records, even of locally abundant taxa, are necessary to clarify this understanding, together with augmenting knowledge of local faunas and diversity.

The major problems of faunal appraisal in Malesia, not in any way restricted to Neuroptera, parallel those noted by Mansell (1990) for African taxa, namely:

1 much type material is old, or has been damaged, lost or destroyed, while most types are housed in institutions far distant from Malesia. This reflects that virtually all species have been described by specialists not resident in the region and, often, with no first-hand knowledge of the insects in the wild or of the habitats from which they come. The types are difficult for local students and researchers to appraise, which is particularly important in attempting to clarify the validity and relationships of the numerous taxa described by L. Navás (the most prolific describer of new Neuroptera over several decades), many of them on small and poorly defined features, and which remain anomalous. For some of these taxa little can be done other than to report the names under which they were described or recorded;

2 most taxonomic work is fragmentary, with many isolated descriptions not fully including comparative information on related taxa, many of which may be masquerading under different generic names. Elucidation of synonymies is necessary to clarify a number of puzzling distributional records, as species have sometimes simply been allocated to the 'nearest available' name;

3 virtually all early research has been on dead adults only, with very little information on early stages and biology;

4 the type localities of many species are given very vaguely on labels; for some species, even the country of origin is uncertain. Several such anomalies are noted later. Referring to Lepidoptera, but reflecting a caveat of much wider relevance, Barlow (1988) wrote 'material collected before 1900 from South East Asia was frequently inaccurately labelled'.

At one level, Neuroptera are an eminently suitable group for a preliminary evaluation such as this. The major family groupings are clear, and the families diagnosed, and many of the genera can be recognised. Most lacewings are relatively large insects, and many are brightly coloured and conspicuous. Studies on the faunas of, particularly, parts of Malaysia, Indonesia and Papua New Guinea have led to clarification of some of the major features of the regional fauna. In contrast, far less information is available from many of the countries to the north and west of Malesia, and the Malesian data are important also in helping to clarify those more distant faunas and their evolution. To the south, the Australian fauna has received considerable attention, and clarifying affinities of that large fauna is also facilitated by study of the Malesian taxa.

chapter **2**

The Neuroptera

Neuroptera is one of the smaller orders of holometabolous insects: about 6000 species have been described (Aspöck & Aspöck, 1999). The families represented in Malesia include about 5400 of these. About 450 species have been recorded from Malesia (see checklist), but the identities of many of these need clarification, with some of the 550 or so names listed of unknown validity.

Morphology

Unifying the morphology of the various lacewing families is difficult. Each family has a characteristic pattern of wing venation, some with idiosyncratic nomenclature for particular structures, and unique sets of genitalic features which - particularly in the male - are hard to homologise across families. Some understanding of these features is necessary for identification and diagnosis, and the various specialised structures are illustrated in each of the family accounts.

For most families, incorporation of genitalic information into generic and species diagnoses is a relatively recent trend, and many taxa have historically been erected solely on small features of wing venation. Many of these characters are indeed valid, but the integrity of some is doubtful and may not heed the extent of individual variability that is now suspected to occur. Genitalic examination is usually necessary for critical identification of species. The major taxonomic characters used for each family are noted in context in the family account. Many of the terms are explained further in a glossary (p. 189), in which some general morphological diagrams of basic structure are also included.

The order Neuroptera can be defined as follows:

Adult

Head moderate to large, orthognathous to (rarely) hypognathous (figs. 1, 2); mouthparts chewing, usually with strong mandibles; antennae usually filiform or moniliform, sometimes clubbed, rarely pectinate, compound eyes large, ocelli usually absent (cf. Osmylidae, fig. 44). Prothorax short, except in taxa with raptorial forelegs (Mantispidae, some Berothidae); legs (fig. 3) usually slender, cursorial, more rarely short and sturdy (Ascalaphidae, some Myrmeleontidae), forelegs rarely raptorial. Pterothorax sturdy; two pairs of subequal membranous wings, hindwings rarely reduced or absent, or greatly elongated and filiform or spatulate (Nemopteridae); wings sometimes patterned, rarely locally thickened, as by small raised domes (nygmata) on the membrane. Venation usually complex (simplified in Coniopterygidae); major

FIGURES 1-3

1 Antlion (Myrmeleontidae), with some major body parts, wing veins and cells indicated. Note that abdominal segments are indicated with Roman figures — **2** head of Neuroptera, with major parts indicated — **3** leg of Neuroptera, with major parts indicated.

1

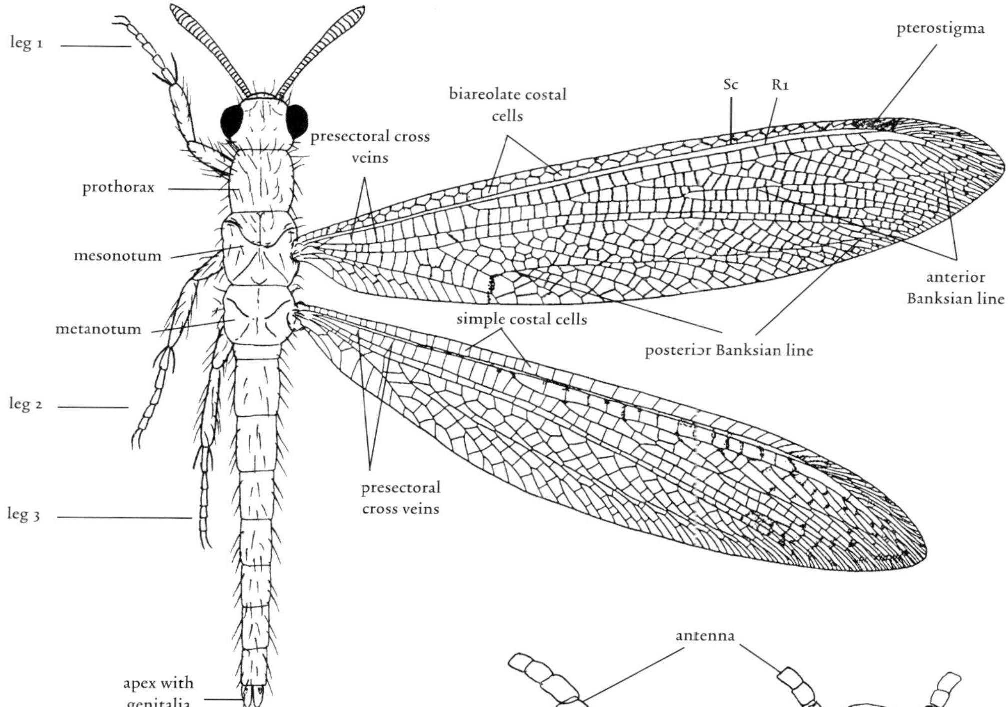

2

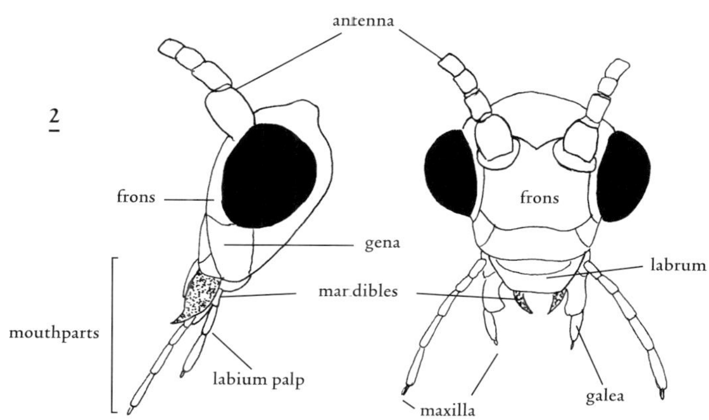

3

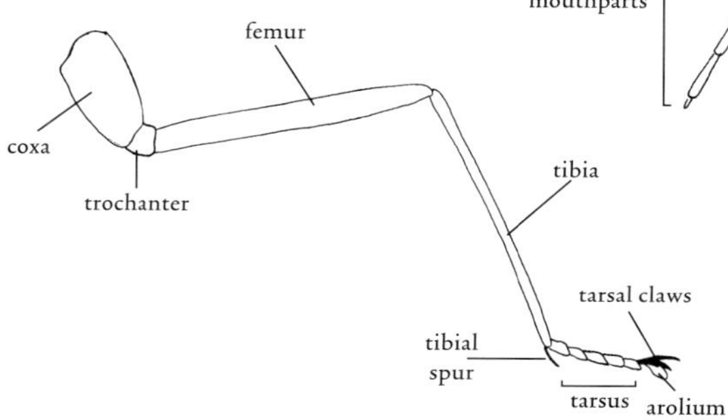

veins often with extensive end-twigging, trichosors often present around margin. Abdomen 10-segmented (nine apparent segments in Chrysopidae); cerci absent, but cercal callus usually present with trichobothria in a clearly defined field. Genitalia complex, and idiosyncratic in many families; female ovipositor exserted only in Dilaridae and some Mantispidae;

Egg

Usually simple elongate ovoids, sometimes with the chorion patterned;

Larva

Mainly terrestrial and freeliving, aquatic (Sisyridae, Nevrorthidae) or semi-aquatic (some Osmylidae), rarely otherwise. Usually active and campodeiform, occasionally scarabaeiform (subterranean Ithonidae) or vermiform (parasitic Mantispidae); prognathous, mouthparts strongly exserted with mandible and maxilla of each side associated closely to constitute a 'sucking tube'; maxillary palpi absent; tarsi one-segmented. Larvae lack an internal connection between the midgut and hindgut, so faecal material can not pass during larval life;

Pupa

Decticous, exarate, usually formed in silken cocoon.

Biological outline

The biology of the Neuroptera, summarised generally by New (1986a, 1989a), reveals a diversity of habits in this widespread order, with many of the families each having characteristic ecological features. The considerable biological variety is noted more specifically in the accounts of the various families, but it is important to emphasise here that much generalisation has resulted from uncritical extrapolation from studies on very few species. For some large groups, the single species studied, even fragmentarily, may not be biologically typical, and far more biological investigation is needed for most lacewing groups. The following brief perspective summarises some of the broader generalities to help establish the basis for the considerable individual variety which occurs.

Eggs are laid singly or in batches, with the particular pattern of deposition highly characteristic for particular taxa. They are usually found on vegetation or on the ground, or flattened on foliage surfaces, but sometimes with long stalks raising the eggs well above the substrate (Chrysopidae, Nymphidae,

Berothidae, Mantispidae). Some Ascalaphidae also produce aborted eggs ('repagula') among batches of normal eggs, which provide wandering predators (such as ants) with distracting food (Henry, 1972). Repagula may also be consumed by the hatchling larvae before they disperse.

Hatching occurs by use of an oviruptor in many families, and otherwise by the larva eclosing through a hole created from a line of weakness in the chorion, so that an end of the egg is pushed off (many Myrmeleontoidea).

Larvae disperse, and those of most families are freeliving predators which feed on a wide variety of small, soft-bodied animals which they pierce with their suctorial mouthparts and extract the body contents. Most species undergo three instar stadia. As predators, larvae may be active and typically campodeiform in appearance (Hemerobiidae, Chrysopidae) or more sedentary 'ambush predators' such as the pit-forming larvae of some antlions (Myrmeleontidae). The more specialised feeding habits include the close association of spongeflies (Sisyridae) with freshwater sponges, the parasitic mode of life of mantid flies (Mantispidae) within spider egg sacs or social Hymenoptera, and the apparent association of beaded lacewings (Berothidae) larvae with termite nests. Larvae of ithonid lacewings (Ithonidae) appear to be particularly unusual in living underground and feeding on plant material.

Pupation takes place within a silken cocoon, the walls of which may also incorporate materials such as sand grains or bark fragments, which increase crypsis. The life cycle ranges from a few weeks to a few years (some Myrmeleontidae), with most species completing at least one generation each year. Well-defined seasonality patterns and phenological breaks, such as overwintering diapause, are widespread in temperate region species, but little information exists on lacewing life history patterns in the tropics.

Adult Neuroptera vary widely in dispersal proclivity. Many are strong flyers, while others appear to disperse little. Different taxa may be predominantly diurnal, crepuscular or nocturnal.

Sex ratios are commonly close to unity. Fecundity can extend as high as several thousand eggs, with the highest numbers reported for Mantispidae. Fewer eggs appears to be more usual: females of many lacewings appear to lay at least 100-300 eggs. Adults of most species are predatory, but others (some Chrysopidae) feed on honeydew and plant exudates, or may be more broadly polyphagous to eat various plant and animal materials; detailed investigation of the diet of most species has not yet been made.

Likewise, within broad constraints of opportunity and physical limitations, the prey spectrum of predatory species may be related to their preferred habitats. Many temperate-region arboreal taxa are found solely or predominantly on either broad-leaved or coniferous trees, and such apparent specificity may

be related to prey choice or suitability. Other kinds of habitat specialisation are numerous and collectively demonstrate the considerable ecological breadth of the order.
Many predators and parasitoids attack Neuroptera, and cannibalism is also common. Some adult antlions (Myrmeleontidae) appear to select other antlions for prey (Stange, 1970), and a variety of other insect predators have been reported incidentally, including asilid flies and sphecoid wasps. Larvae of antlions are attacked by those of bombyliid flies. Parasitoids of many lacewings are known, and any stage may be attacked. There are few relatively comprehensive surveys, but a variety of Proctotrupoidea, Chalcidoidea and Ichneumonoidea have been recorded. Records from Malesia are very sparse, and more information from the region will be of considerable value in augmenting current knowledge derived mainly from temperate regions.

Higher classification

Seventeen recent families of Neuroptera occur (or eighteen if the separation of Berothidae and Rhachiberothidae is accepted). However, the phylogenetic relationships between these are still debated, although there is little doubt that the order itself is monophyletic. This presumption is derived largely from the unifying larval characters, particularly the suctorial jaws and lack of internal connection between midgut and hindgut. Unifying synapomorphies for adults are more obscure.
For many decades, an arrangement of six major groupings of Neuroptera has been followed by many workers (table 1; these groups are widely considered as superfamilies, but with some families being moved between some of these at intervals). Recently, Aspöck (1992), followed by Aspöck et al. (2001), has provided evidence for three major natural lineages (fig. 4). The latter scheme recognises the isolated position of Nevrorthidae, in which the larva has a well-developed gula. The remaining families are divided among two major groupings, essentially comprising the Myrmeleontoidea of most previous authors, and the rest. The last group thereby associates many of the families regarded as more distantly related by earlier students. Thus, Coniopterygidae (which differ considerably in appearance from all other lacewings) and Ithonidae have each been regarded widely as isolated and primitive 'off-shoots' from the main development of the order. Debate over the affinities between some families will continue, but it is virtually certain that all major groups of the Neuroptera are now known.

The sequence of family treatments in this book largely follows the more traditional arrangement of table 1, other than in allying the Dilaridae more closely with Mantispoidea than with Osmyloidea.

Several families of Neuroptera are absent from Malesia, or represented only very poorly in the region. Polystoechotidae is a small New World family. Nevrorthidae (formerly Neurorthidae) are highly disjunct, with representatives in southern Europe, Taiwan, Japan and eastern Australia; they could, possibly, be found in New Guinea streams. Nemopteridae are recorded only by an unconfirmed early record from Java and may not be present. Van der Weele (1909) noted that a purported nemopterid described as *Himanopterus fuscinervis* Wesmael is a lepidopteran, and his forecast that the family would be found in the region has not yet been fulfilled. Ithonidae sens. rest. (that is, as a group distinct from Rapismatidae, see p. 36) have not been reported. Psychopsidae, known from Australia and parts of Southeast Asia (southern China, Myanmar, Laos, Taiwan) also appear to be absent, and it is unlikely that such spectacular insects would have escaped notice in Malesia until now. Most other families have been reported, although some of them only rarely and from few species on the fringe of a larger geographical range.

TABLE 1
Major groupings of families of Neuroptera into superfamilies. Compare with fig. 4. See text for comments.

ITHONOIDEA	CONIOPTERYGOIDEA	OSMYLOIDEA
Ithonidae Rapismatidae	Coniopterygidae	Osmylidae Sisyridae Nevrorthidae
MANTISPOIDEA	**HEMEROBIOIDEA**	**MYRMELEONTOIDEA**
Mantispidae Berothidae Rhachiberothidae Dilaridae	Hemerobiidae Chrysopidae Polystoechotidae Psychopsidae	Nymphidae Nemopteridae Ascalaphidae Myrmeleontidae

FIGURE 4
Higher classification of Neuroptera.
a Three major lineages of Neuroptera (after Aspöck, 1992). Compare with the more traditional arrangement shown in table 1 — **b** results of a cladistic analysis of relationships of the families of Neuroptera; the 'preferred cladogram' of Aspöck et al. (2001)

a

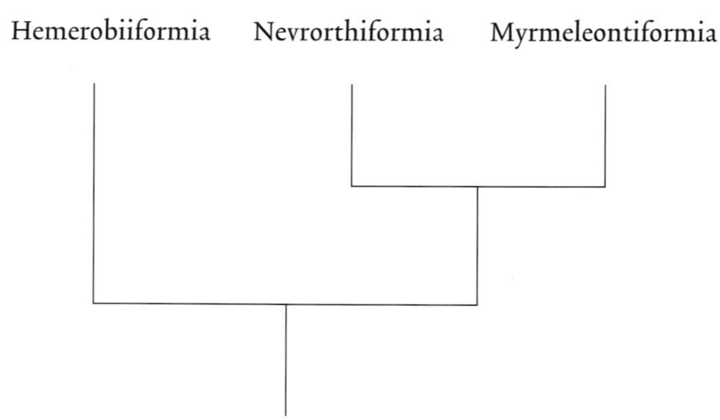

b

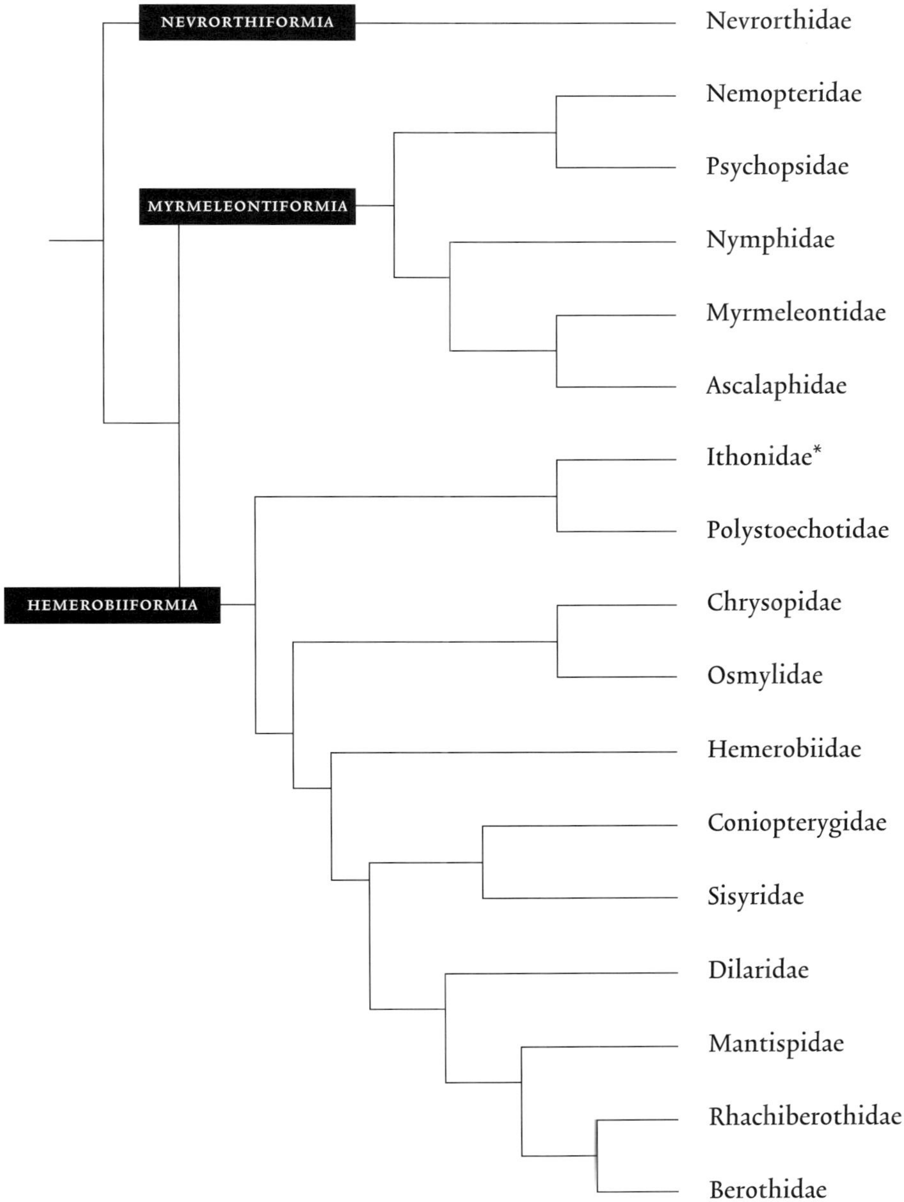

* includes Rapismatidae (see text)

Key to the families

Because of uncertainties over some family incidences in Malesia, all extant families of Neuroptera are included in the following keys. The keys do not heed any formal superfamily groupings, other than by coincidence, as they are designed primarily for practical recognition and identification of specimens. Representative adult lacewings are shown under the respective families and illustrations are provided in context to the features of all families. Each family is also diagnosed more fully in the systematic account (with some details relevant only to extralimital taxa omitted), and decisions made from this key can thus be augmented.

Family key for adult Neuroptera (after New, 1989a)

1 Small, forewings to about 5 mm, usually smaller; wings and body covered with white or greyish waxy or mealy secretion; costal area with no more than one or two (basal) crossveins (fig. 19) ▸ **Coniopterygidae** (p. 39)

– Not as above: larger, not waxy, costal area with more numerous crossveins ▸ 2

2 Forewings and hindwings of markedly different shape: hindwing greatly elongated, filiform or spatulate ▸ **Nemopteridae** (not in Malesia?)

– Forewings and hindwings generally similar in shape, though hindwing may be shorter or smaller than forewings (or, more rarely, larger) ▸ 3

3 Forelegs raptorial (fig. 54), with femur strongly thickened; prothorax elongate ▸ 4

– Forelegs not raptorial; prothorax usually short ▸ 5

4 Forewing costal crossveins simple; trichosors absent (figs. 62, 63) ▸ **Mantispidae** (p. 59)

– Forewing costal crossveins often forked; trichosors present (fig. 68) ▸ **Berothidae**: Rhachiberothinae (or Rhachiberothidae, not in Malesia)

5 Antennae filiform or moniliform, rarely pectinate (male Dilaridae), never clubbed or with apex conspicuously broadened ▸ 6

– Antennae with distinct club or other apical/preapical thickening ▸ 17

6 Stout bodied; hindwings very broad near base; head often partially retracted under pronotum ▸ 7

– Not as above; if, rarely, with body stout, hindwings no broader than forewing at base ▸ 8

7 Tibial spurs absent (fig. 5); claws broad, with small subapical projection; trichosors weakly defined ▸ **Rapismatidae** (p. 36)

– Tibial spurs present (fig. 6); claws elongate; trichosors well developed ▸ **Ithonidae** (not in Malesia)

8 Apparent ocelli present as three raised tubercles on vertex (fig. 44) ▸ **Osmylidae** (p. 53)

– No trace of ocelli, although small vertex tubercles present in some Psychopsidae and Dilaridae ▸ 9

9 Forewing vein Sc parallel with R1 and fused distally with this and with Rs to form a 'vena triplica' (fig. 7). (Wings very broad; Rs with numerous branches, crossveins in defined gradate series.) ▸ **Psychopsidae** (not in Malesia)

– Not as above; no vena triplica and not with above combination of secondary features ▸ 10

10 Veins Sc and R1 fused near apex of wings (fig. 126) ▸ 11

– Veins Sc and R1 not fused near apex of wings ▸ 12

11 Both wings with two defined series of gradate crossveins; few R1-Rs crossveins; recurrent humeral vein (fig. 8) present ▸ **Polystoechotidae** (not in Malesia)

– Both wings with numerous irregular crossveins, including many between R1 and Rs; no recurrent humeral vein (fig. 9) ▸ **Nymphidae** (p. 107)

12 Veins Sc and R1 separate apically, not connected by crossvein (fig. 76) (male: antennae pectinate; female: ovipositor exserted); wings usually densely hairy ▸ **Dilaridae** (p. 72)

– Veins Sc and R1 connected near apex (fig. 91; antennae simple, ovipositor not exserted) ▸ 13

13 Forewings with at least two apparent Rs veins branching from R1 (figs. 89, 91, 94, 95) ▸ **Hemerobiidae** (p. 75)

– Forewings with only a single Rs branching from R1 ▸ 14

FIGURES 5 - 9
Neuropteron legs without **5** and with **6** tibial spur — **7** vena triplica in forewing of representative psychopsid — **8** presence, and **9** absence of recurrent humeral vein at base of forewing costal cell.

5

6

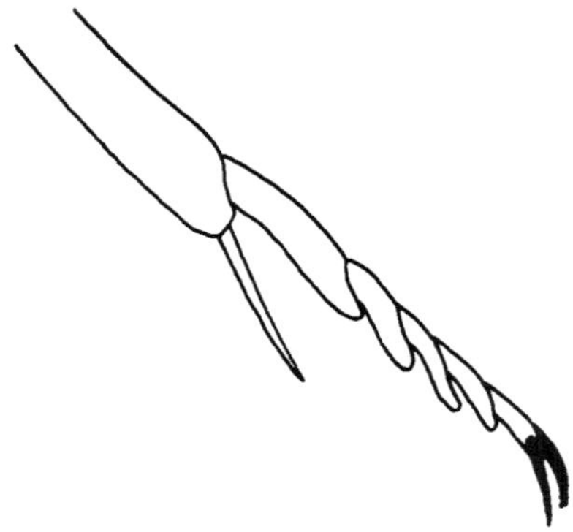

7

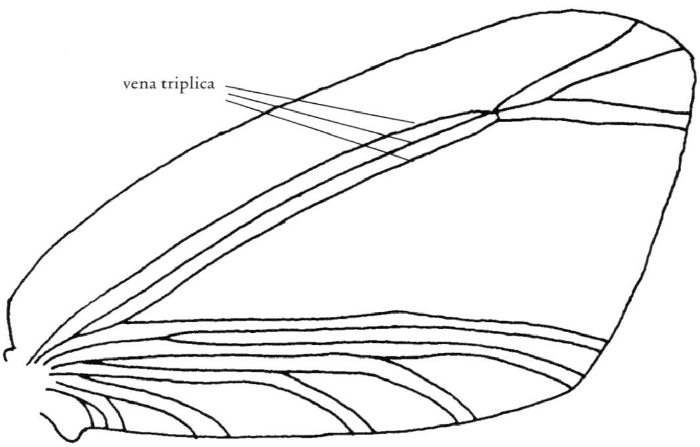

8

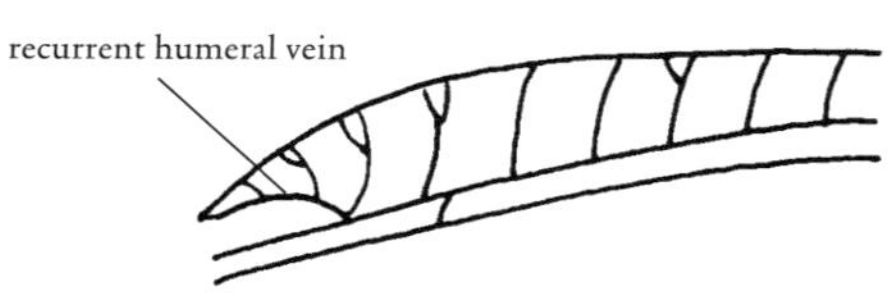

9

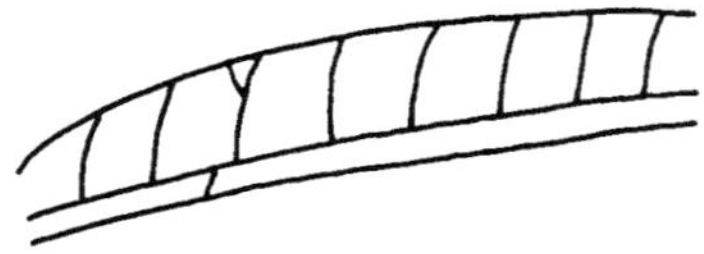

14 Hindwing vein CuA (fig. 42) not conspicuously close to hind margin (medium to large, usually green, trichosors absent) ▸ **Chrysopidae** (p. 50)

– Hindwing vein CuA close to hind margin for some distance (often small, rarely green, trichosors may be present) ▸ 15

15 Forewing costal cross veins usually all simple (fig. 42) ▸ **Sisyridae** (p. 68)

– Forewing with at least some costal crossveins forked ▸ 16

16 Forewings with several Sc-R1 crossveins; wings not hairy ▸ **Nevrorthidae** (not in Malesia?)

– Forewings usually with only a basal Sc-R1 crossvein (fig. 70), rarely a second crossvein at or beyond half length of wing; wings often densely hairy or with flattened scales on veins ▸ **Berothidae** (excluding Rhachiberothidae; p. 68)

17 Antennae longer than half length of forewing, strongly clubbed ▸ **Ascalaphidae** (p. 111)

– Antennae usually much shorter than half forewing length, more or less gradually expanded or clubbed at apex ▸ **Myrmeleontidae** (p. 122)

Family key for late instar larvae (after New, 1989a)

NOTE
The probable appearance of larvae of Dilaridae, Berothidae and Nymphidae to be found in Malesia is shown under the respective familly accounts by representatives from other parts of the world.

1 Abdominal segments with paired ventral gills (aquatic); labial palpi absent (figs. 39, 43) ▸ **Sisyridae** (p. 50)

– No ventral abdominal gills; labial palpi present ▸ 2

2 Labial palpi two-segmented, with apical segment enlarged; antennae two-segmented (larvae small, usually only 2-4 mm; fig. 20) ▸ **Coniopterygidae** (p. 39)

– Labial palpi and antennae both with more than two segments (larvae larger) ▸ 3

3 Mandibles with large tooth or teeth along inner edge (figs. 150, 172) ▸ 4

– Mandibles without large tooth or teeth along inner edge (figs. 45, 81, 107, 218), occasionally with small papillae but usually smooth ▸ 6

4 Mandibles each with one preapical tooth; hind tibiae and tarsi not fused ▸ **Nymphidae** (p. 107)

– Mandibles each usually with more than one (most commonly, three) preapical tooth; hind tibiae and tarsi usually fused ▸ 5

5 Lateral body processes (scoli) well developed, usually two each on mesothorax and metathorax and one each on first seven abdominal segments; hind margin of head usually strongly excavated; hind tarsal claws normal (fig. 150) ▸ **Ascalaphidae** (p. 111)

– Lateral body processes absent or weakly developed; hind margin of head usually not excavated; hind tarsal claws often enlarged and flexed forward (fig. 172) ▸ **Myrmeleontidae** (p. 122)

NOTE

Larvae of the two families keyed here are sometimes very similar, and their body forms intergrade considerably. The features given are based on few larvae known from Malesia.

6 Either neck elongated with prothorax long and slender or mandibles with several papillae along inner edge ▸ **Nemopteridae** (not in Malesia)

– Not as above ▸ 7

7 Body scarabaeiform; tibiae and tarsi fused; eyes absent ▸ **Ithonidae** (not in Malesia)

– Body not scarabaeiform; tibiae and tarsi separate; eyes present ▸ 8

NOTE

The unknown larvae of Rapismatidae may key to this couplet and cause confusion.

8 Tarsi with one claw; mandibles short (fig. 55; often in spider egg sacs) ▸ **Mantispidae** (p. 59)

– Tarsi with two claws; mandibles long (free-living) ▸ 9

9 Mandibles and maxillae inwardly curved from base to converge toward apex, often broad based ▸ 10

– Mandibles and maxillae straight, usually slender; occasionally incurved at apex ▸ 12

FIGUUR 9a
Larval Nevrorthidae

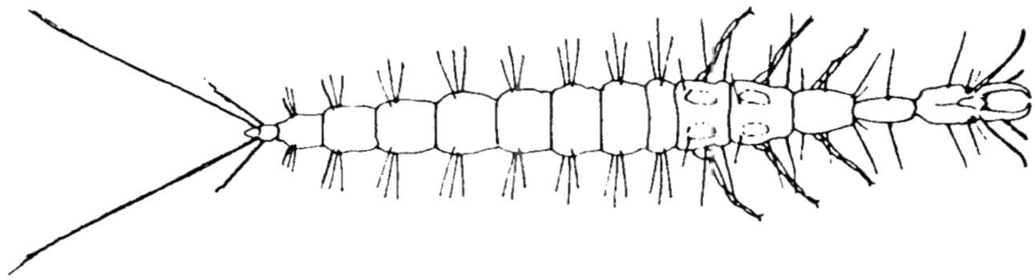

10 Labial palpi five-segmented; antennae ten-segmented; frons tapered anteriorly ▸ **Psychopsidae** (not in Malesia)

– Labial palpi with more than five segments; antennae with one segment very long and with secondary annulations giving impression of numerous ill-defined segments; frons not tapering ▸ 11

11 Empodium trumpet-shaped; abdomen and thorax often with long hairs, sometimes grouped on protruding tubercles (fig. 107) ▸ **Chrysopidae** (p. 86)

– Empodium not trumpet-shaped (except in first instar); abdomen and thorax with only short hairs, sometimes almost naked (fig. 81) ▸ **Hemerobiidae** (p. 75)

12 Mandibles and maxillae abruptly inturned at tip (aquatic; fig. 9a) ▸ **Nevrorthidae** (not in Malesia?)

– Mandibles and maxillae straight ▸ 13

13 Mouthparts considerably longer than palpi or antennae; antennae multi-segmented or with strong secondary annulations (fig. 45) ▸ **Osmylidae** (p. 53)

– Mouthparts of similar length to palpi and antennae; antennae five to eight-segmented ▸ 14

14 Abdomen very long, most segments longer than broad; labial palpi five to eight-segmented ▸ **Dilaridae** (p. 72)

– Abdomen long, but segments usually broader than long; labial palpi five-segmented ▸ **Berothidae** (p. 68)

Collecting and preparation

Beating and sweeping vegetation are standard collection methods for lacewing adults and larvae associated with vegetation. Others may be found by direct searching, with experience helping to increase the effectiveness of this considerably; thus, Stange & Miller (1990) noted the high incidence of larvae of some antlion groups in sand or loose soil, including that under rocky overhangs, whilst others live on bare rock faces or in caves. Larvae of many taxa can be reared without difficulty, if kept in individual small containers (to avoid losses from cannibalism) and provided with suitable prey and shelter. For arboreal taxa, small soft-bodied Hemiptera (such as aphids) are often suitable prey; many ground-dwelling forms will take ants or various insect larvae.
Adults of many Neuroptera come to light, and light trapping is perhaps the most effective general method for collecting these. Some diurnal taxa, such as many Ascalaphidae, can be netted directly, and nocturnal species may be found at rest on bark or under overhangs during the day. Water traps and other general collecting methods such as Malaise traps also yield lacewings.

Adults can be preserved dry or in alcohol. The latter can lead rapidly to loss of colour and, in many cases, to some deterioration. Other than for the tiny Coniopterygidae (which can be card-pointed) adults are pinned vertically through the mesothorax. Because of the importance of wing venation in identification, the wings (at least of one side of representative specimens in a series) should be set to facilitate examination. Many workers prefer to spread wings of both sides, as for Lepidoptera. All specimens should be labelled fully in standard manner.

Genitalic examination is also critical for species diagnosis and, in some broader instances such as genera of Chrysopidae, identification can be done only after clearing the abdominal apex and close examination of small genitalic elements. Methods for doing this satisfactorily vary somewhat in the protocols of different workers, but generally involve the sequence of (1) removing the abdominal apex by cutting carefully with fine scissors (wetting the blades with water before use will reduce the chances of loss of dry cut pieces by 'catapulting'), (2) macerating this in 5-10% sodium or potassium hydroxide, either hot (for 5-10 minutes) or cold (several hours to overnight) to remove muscular tissue, (3) washing, and if necessary neutralising the alkaline action, (4) if necessary, staining with dilute aqueous chlorazol black 'e' or acid fuchsin (staining may not be necessary for routine examination, but is needed if critical examination of fine details of ducts is necessary), and (5) dehydrating to store in alcohol, glycerine, or to slide mount. The latter is usual for very small specimens, such as all Coniopterygidae, so that the various minute parts do not become lost after separation and dissection, but storage of the whole cleared abdominal apex in a microvial of liquid (such as glycerine) associated with the parent specimen is also common. The microvial can be attached to the same pin as the donor, or on a separate pin, with both parts then labelled so that the association is unambiguous. For critical examination of delicate structures, such as the fine ducts of various female Chrysopidae, Adams & Penny (1987) recommended study 'in the half shell', after cutting the abdominal plates along the insect's midline (using fine iris scissors) and removing those of one side to leave the internal structures in situ. Changes in the genitalia of both sexes may occur with maturation, with increased sclerotisation and development of structures. Adams & Penny (1987) noted that dissection of teneral (very young, newly emerged) specimens of Chrysopidae could provide misleading information.

Measurements of body length are made from the front of the head to the tip of the extended abdomen. Wing length is the greatest length, in a straight line from base to apex.

Economic importance of Neuroptera

The major - and, certainly, most familiar - interest to many entomologists is the widespread predatory habit of larvae and adults of Neuroptera, particularly those living on vegetation, which gives them importance in the biological control of arthropod pests on crops. This interest is most developed for some members of Chrysopidae and Hemerobiidae, with a relatively minor applied role for the Coniopterygidae. The biocontrol values of most other groups are currently small. Not surprisingly, far more is known of the biology of Chrysopidae and Hemerobiidae than of any other family, largely as a consequence of attempts to mass-rear and manipulate species as predators in the northern temperate regions (McEwen et al., 2001). Several species of Chrysopidae, in particular, have been studied in great detail and are used commonly in management of pests on field crops. The genus *Chrysoperla* Steinmann is paramount in this context (Brooks, 1994, McEwen et al., 2001).

chapter 3

Biogeography

The biogeographical region known as Malesia is primarily a botanical construct based on the floristic region which comprises Indonesia, West and East Malaysia, southern Thailand, the Philippines and Papua New Guinea (fig. 10). To the north and west it is taken to extend to the Isthmus of Kra, which separates southern Thailand and peninsular Malaysia from northern Thailand, Myanmar and adjacent countries, and indicates a major floristic disruption between Malesian and continental Asian floras. The eastern boundary is less easily defined, with different authorities including (Kato, 1989, following Van Steenis, 1950) or excluding (Laumonier, 1997) New Britain and nearby islands to the east and southeast of the main landmass of New Guinea. Whitmore (1987) noted that the region extends 'to New Britain' but without clear indication over whether it was included. The region (including New Britain) thereby extends over nearly 60° of longitude (approximately 95-153° E, about 6600 km) and nearly 30° of latitude (approximately 18° N - 11° S, about 3400 km), and is biogeographically complex as a meeting place of the Asian and Australasian biotas. In Malesia there are 12 Neuroptera families with about 450 species.

Malesia

Following Van Steenis (1950), many botanists recognise three major subdivisions of Malesia: East Malesia, South Malesia and West Malesia (fig. 13). The east and west subregions are totally distinct, each with many endemic plant genera. In contrast, Java and the Lesser Sunda islands have relatively poor flora, but this constitutes a characteristic 'monsoon flora' (Laumonier, 1997). For fauna, the region's significance has long been recognised, with lines of faunal demarcation such as Wallace's line, Weber's line and Lydekker's line indicating some of the interpretations of faunal intergradation (fig. 12). Wallace's line was designed to denote the boundary of Asia, and many authors have since debated where this should properly be sited (for example Simpson, 1977). As MacKinnon & MacKinnon (1986) commented, the problem is of immense complexity in trying to find the most effective and acceptable compromise between the distribution patterns of many different groups of animals with different rates of evolution and widely varying powers of dispersal. The MacKinnons noted the need to find 'lines of balance', marking the regions where opposite influences balance, and 'lines of sudden change', delimiting distinctive changes in species composition corresponding to natural isolating barriers. Wallace's line (as modified by Huxley, 1868) follows the edge of the Sunda shelf; Lydekker's line is defined by the edge of the Sahul shelf (and thus defines the true boundary of the Papua-Australian region), and Weber's line denotes faunal balance between these.

FIGURE 10
Map of Malesia (black) with major political entities named.

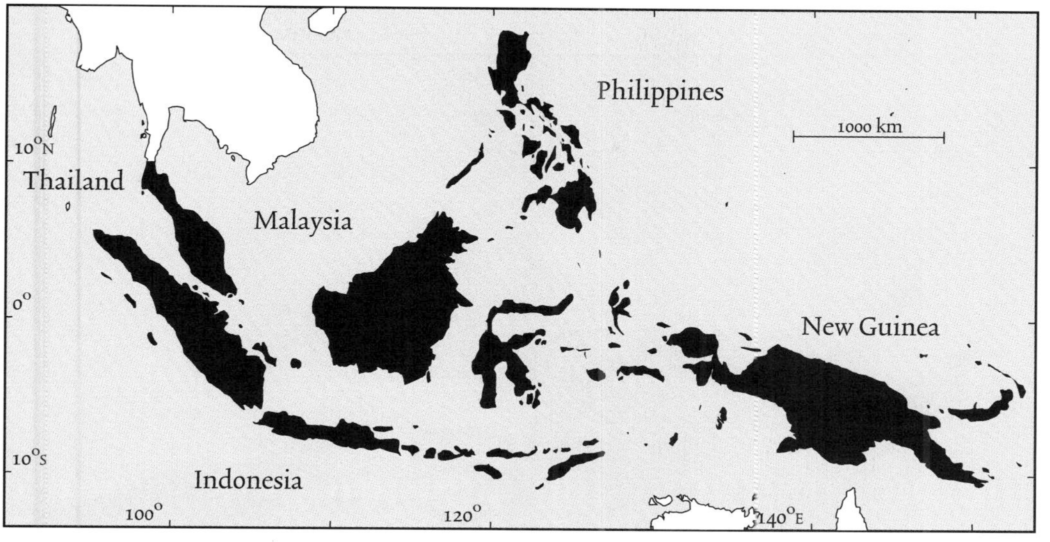

FIGURE 11
Wallacea, with the major regions indicated.

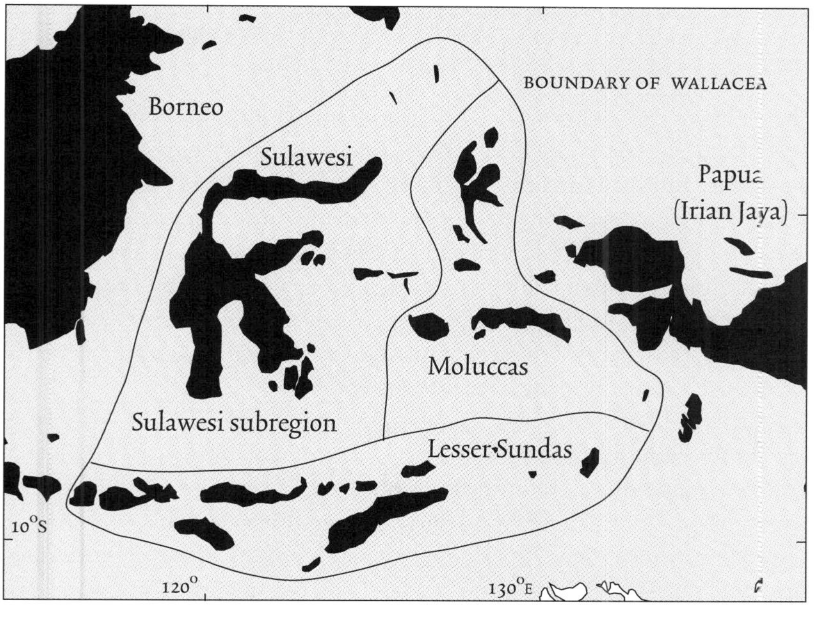

FIGURE 12
Major faunal subdivisions of Malesia, indicating positions of Sunda shelf and Sahul shelf (dark grey) and some of the divisions proposed as biotic boundaries in the region (see text). The dates marked on Wallace's line show the boundaries adopted or recommended at different times.

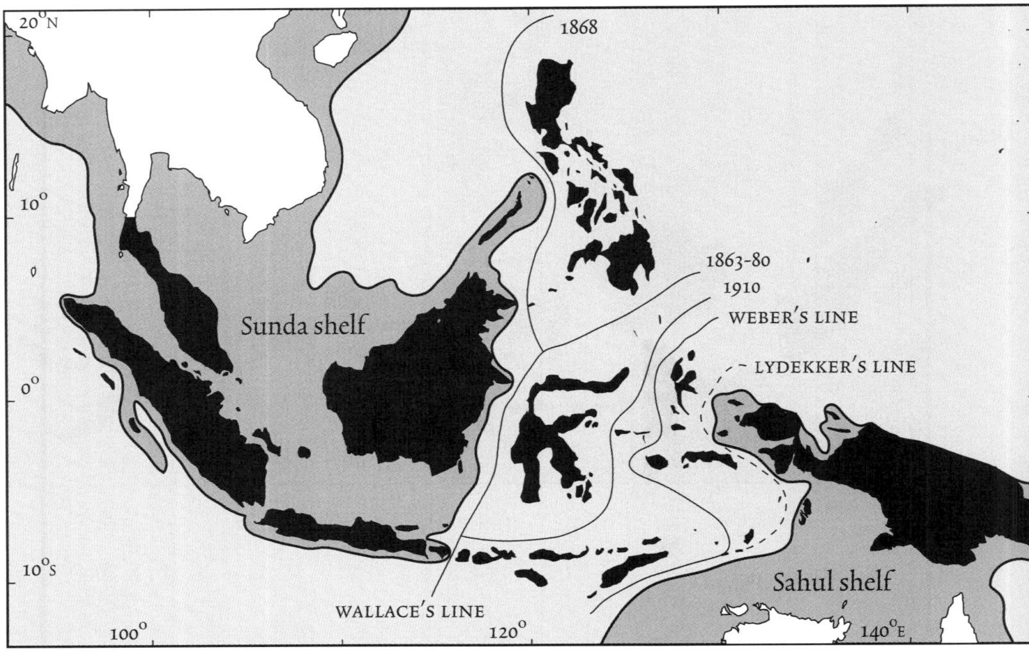

The region to the west of Wallace's line is often delimited as the Sundaic subregion, and that to the east of Lydekker's line as the Sahul subregion. Between the Wallace and Lydekker lines, the transitional area known as Wallacea (fig. 11) has an essentially Asian flora, but the fauna is a mixture of Asian and Australian elements, with considerable local endemism on the complex archipelagos present. Additional faunal subregions have been suggested. The integrity and boundaries of Wallacea are debatable, with considerable discussion since Dickersen et al. (1928) defined the area to include the Philippines, a region which is considered distinct by many later authorities.

Although a discussion of the biogeographical complexities of Malesia is beyond the scope of this handbook, the variety of possible faunal origins and affinities is indeed important in attempting to interpret the evolution of any taxonomic group, and in defining the scope of a treatment of relatively poorly known organisms. Thus, in contrast to some groups of larger butterflies (Ackery & Vane-Wright, 1984) and cicadas (Duffels & de Boer, 1990), which are

FIGURE 13
Botanical subdivisions of Malesia (after Van Steenis, 1950).

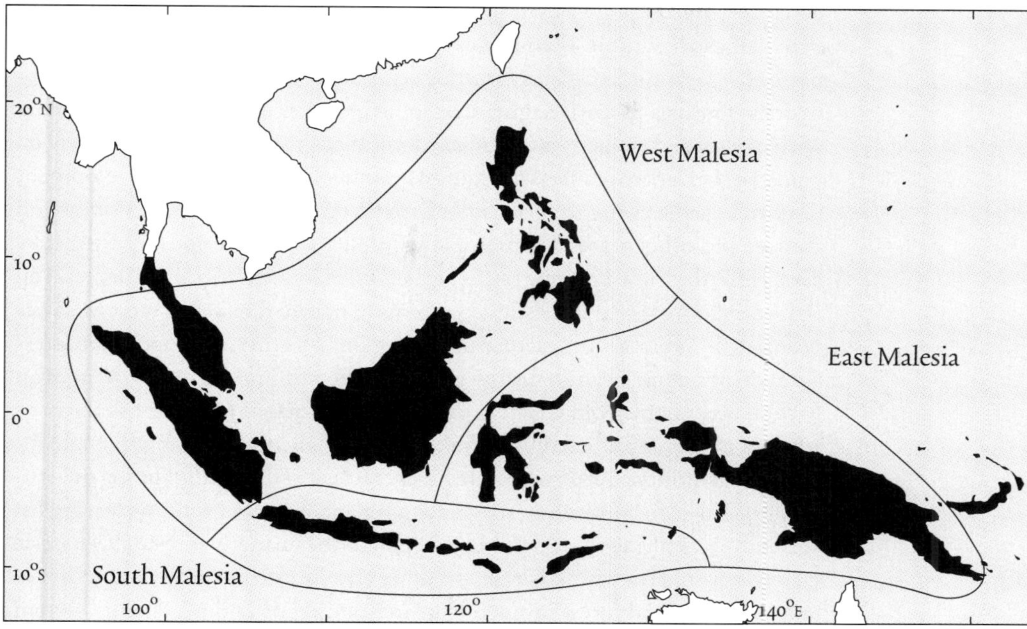

reasonably well documented and for which a number of areas of endemism can be indicated reliably, appraisal of lacewings necessitates also some more detailed consideration of the faunas of regions abutting Malesia. Thus, I am including also comments on the Neuroptera of nearby regions, and indicate which of the elements not yet found in Malesia may be expected to be revealed as collecting becomes more intensive.

However, there is an important caveat to this, namely that such collecting in natural habitats may never be possible over much of Malesia. Many parts of Malesia have undergone dramatic change over the last century or so, and continue to do so to accommodate the needs of massive human population increase. Many of the unusual lacewings were described before the period of major economic expansion, based on tin and rubber, which resulted from colonial rule in Malaysia for example (Aiken, 1994), which was associated with deforestation and landscape degradation. Increased intensity and extent of agriculture and settlement has inevitably led to large scale clearance of native vegetation and its replacement by crops, often accompanied by high input of agricultural chemicals. Indeed, referring to the whole of the Indo-Malayan region, MacKinnon & MacKinnon (1986) noted that human population pressures on land are higher than in any other major region on earth. The Malay Peninsula, for example, then had only around 40% of its area still covered with natural vegetation, mostly as forest reserves for hydrological purposes or production forestry. Intensive agriculture of irrigated rice is widespread, together with extensive development of plantations of economic crops, particularly oil palm, rubber, cocoa, coffee, coconuts. As elsewhere in the region, development of tourist facilities is also increasing, with accompanying construction of golf courses and other resort facilities. Many localities made famous for insects by early collectors now bear few traces of their former inspirational character. Yet, the reservation of characteristic ecosystems in nature reserves in West Malaysia is amongst the most satisfactory in the region. The massive population pressures on Sumatra, Java and nearby islands to the east impose severe changes on natural ecosystems, and effective protection and management for designated reserves is often minimal: it was ranked as 'poor' to 'non-existent' by MacKinnon & MacKinnon (1986), who noted 'general lack of discipline among government officials' in addition to illegal logging and other vegetation clearing for agriculture as prime threats to biota. Much of the formerly forested Indonesia is now not so: Java with less than 9% of its area forested in 1986, and much of Sumatran lowland forest invaded by plantation crops under a 'kebun' system of cultivation. Central islands of the Philippines are reported to be 'almost totally stripped of forest' (MacKinnon & MacKinnon, 1986). Palawan is the only reasonably extensively forested island, reflecting relatively low human population density (MacKinnon, 1997). However, Palawan is abundantly distinct from the remainder of the Philippines by being on the Sunda Shelf (fig. 12).

In such climates of intensive human needs, it is understandable that conservation of poorly known native biota and habitats has low priority, and much of the fauna will increasingly depend on limited areas of natural vegetation and the harmony of the intervening land matrix. Such 'intermediate' habitats are critical for much local wildlife, but their increasing frequency implies, as for many other groups of organisms, that loss of lacewing species which depend on more specialised habitats in the region is likely to have occurred. Several of the type localities of Rapismatidae collected many years ago, for example, are now largely occupied by plantation crops (such as oil palm) with high pesticide inputs and consequent degradation of nearby areas.

At present, the Philippines are probably subject to higher levels of irreversible biodiversity loss than any other country in the region (MacKinnon, 1997). In short, the statement by Braatz (1992) that 'Unless immediate, decisive steps are taken to counter the effects of deforestation and other forms of natural resource destruction [...] much of Asia's biodiversity will be irreversibly lost within this generation' is entirely realistic, and ecologically specialised Neuroptera may already be affected. Updating the MacKinnons' (1986) review, MacKinnon (1997) noted that in the Indo-Malayan region as a whole 'the overall situation for biodiversity conservation looks much worse than it looked ten years ago'.

Faunal analysis

The primary theme of this account is to evaluate the extent to which Malesia is a region of faunal mixing and interchange between the Neuroptera of the Australian and Oriental/eastern Palaearctic regions, as an aid to further defining these parent faunas. The limited data available, and the great geographical complexity of Central Malesia, renders most inferences tentative.

As noted earlier, few local faunas in Malesia have been evaluated in detail, not least because of a dearth of resident neuropterists and purposeful surveys in the region, so that most species have been described from more or less incidental collections and a small number of captures made by entomologists interested primarily in other taxa and, in many instances, from only short visits to any particular locality. For example, the capture of a specimen of *Isoscelipteron nicobaricum* on the Krakatau Islands, Indonesia, extended the known range of this berothid considerably (New & Sudarman, 1988), and Monserrat's (1981) captures of Coniopterygidae provided some notable range extensions in Malaysia. Local diversity of lacewings in Malesia is very unevenly documented, and there have been few attempts at any comprehensive survey, even as natural enemies of crop pests. Thus, Dammerman (1929) noted only three species of lacewings in association with crop pests on the Malay Peninsula. It is likely that many species are far more widely distributed than the present, often highly disjunct, records imply. Many of the tantalising early records of species noted in context earlier and not confirmed by recent collecting are thus difficult to dismiss, and collectively indicate the great difficulty in defining the fauna fully.

New (1989c) summarised the results of two periods of fieldwork (totalling nearly ten weeks) of collecting lacewings by a variety of methods around Wau

(Morobe Province, Papua New Guinea), building on collections made over a long period at the Wau Ecology Institute to constitute samples from the most intensively investigated area in the country. More than 65% (41/61) of species known from the area were collected during this relatively intensive survey, and the more extensive Wau collections have provided about a third of all species collected from Papua New Guinea. Many of the species collected at Wau are still known from only very small numbers, and the collections are not in any way fully definitive for the fauna. Likewise, surveys of the small Krakatau islands, Indonesia (New & Sudarman, 1988; Tsukagushi & Yukawa, 1988) led to implications of occurrence of widespread taxa with high vagility, but also included previously undescribed taxa found only in one or other survey. The Krakatau studies are of particular importance in demonstrating that some taxa of lacewings not normally thought to disperse readily (such as *Isoscelipteron* and *Spilosmylus*) may colonise areas far from previous habitations.

The richness of the families of Neuroptera reported from Malesia is summarised in table 2, in which the constitution of the known global fauna is also given for comparison. Globally, Myrmeleontidae and Chrysopidae are by far the most speciose families of lacewings, and there is then a substantial gap to

TABLE 2
Summary of species richness (numbers of described species only) of families of Neuroptera found in Malesia, with indication of global diversity. The six non-Malesian families (table 1) are omitted. The number of taxa of each family known from single incidences or single localities in Malesia is also given.

FAMILY	MALESIA TOTAL NUMBER	MALESIA SINGLE RECORD/SITE	WORLD*
Rapismatidae	11	7	15
Coniopterygidae	59	42	430
Sisyridae	3	1	50
Osmylidae	56	37	150
Mantispidae	63	35	420
Berothidae	6	3	80
Dilaridae	3	3	40
Hemerobiidae	50	24	550
Chrysopidae	105	53	1250
Nymphidae	9	5	35
Ascalaphidae	25	9	380
Myrmeleontidae	60	23	2000
Total	450	242	5400

* These figures are very approximate because of numerous undetected synonymies in some larger groups, but indicate the relative richness of the different families. Many of these figures are based on those given by Aspöck & Aspöck (1999).

TABLE 3

Malesian Neuroptera: numbers of species recorded from major countries in the region, and main faunal subregions: A = Sundaic subregion, B = Wallacea, C = Sahul subregion.

FAMILY	NUMBER OF FORMALLY DESCRIBED SPECIES								
	Thailand	Malaysia	Singapore	Indonesia	Philippines	Papua New Guinea	SUBREGION A	B	C
Rapismatidae	1	7	-	2	-	-	9	-	-
Coniopterygidae	1	13	4	21	5	25	26	7	33
Sisyridae	2	1	-	-	1	-	2	1	-
Osmylidae	2	5	-	29	7	14	12	8	36
Mantispidae	1	10	3	41	11	9	31	27	13
Berothidae	-	2	1	2	1	1	4	1	1
Dilaridae	1	2	-	-	-	-	3	-	-
Hemerobiidae	1	4	-	16	19	15	15	19	17
Chrysopidae	3	23	7	47	25	25	57	34	27
Nymphidae	-	-	-	4	1	5	-	1	7
Ascalaphidae	2	4	-	12	5	3	16	7	3
Myrmeleontidae	1	12	-	18	8	19	19	16	21
Total	15	83	15	192	83	116	194	121	158

the group of families which each contain a few hundred species. Other families are considerably smaller. The Malesian fauna are enumerated by country in table 3, and by the major faunal regions in the area (fig. 12). However, as a mark of 'confidence' in the information, the numbers of taxa known only from single specimens or localities are also noted (table 2), to emphasise the very fragmented and incomplete nature of the template available for extrapolation. The true distribution of these species is thereby unknown, but is almost certainly greater than current knowledge implies; figures of local diversity must be regarded as very tentative. Few taxa are known to be widespread and common in the region, and these are generally ecological generalists known to be polyphagous and vagile, and to tolerate anthropogenic changes. As examples, *Plesiochrysa ramburi*, *Micromus timidus* occur on a wide range of vegetation types, including crops, and their larvae are polyphagous, and *Myrmeleon acer* s.l. can occur virtually wherever soft, sandy soils in sheltered environments are found. The population of *Myrmeleon celebensis* studied by Kiew (1970a, b) was under a building in Selangor, and Kiew (1970a) suggested that ground underneath raised housing has become a major habitat for this antlion. The apparent recent colonisation of Anak Krakatau Islands by *M. frontalis* (New & Sudarman, 1988; Turner, 1992) attests to its ability to find and exploit such environments.

The Australian fauna is diverse (table 4) and contains a distinctive northern element with close affinities to that of Papua New Guinea, in particular. Indeed, Australia and New Guinea have been separated for only some 7000 years by the opening of Torres Strait. Not surprisingly, therefore, the fauna of Cape York has much in common - even at the species level - with that of Papua New Guinea, and the full extent of these similarities has yet to be clarified (New, 1990b). Distributional paths can be postulated in both directions. Thus, Spilosmylinae are very diverse in New Guinea (and Indonesia), but sparse in northern Australia and absent from the rest of the continent, whereas Myrmeleontidae such as *Periclystus* are most typically Australian but extend to the southern fringe of New Guinea. Other groups are more evenly balanced and predominant; thus, *Heteroconis* is by far the predominant genus of Coniopterygidae in the region, both in Malesia and over much of Australia. Nymphidae are most diverse in Australia, but *Osmylops* and *Myiodactylus* are also characteristic New Guinea taxa. Many characteristic Australian lacewing taxa therefore extend into New Guinea and, in some cases, further into Malesia. Christmas Island, although politically Australian, is biogeographically Indonesian. The island has so far yielded only about six species of lacewings, most of them widespread (New, 1991b). No endemics are recognised.

The following Australian families and subfamilies are not known to occur in Malesia: Ithonidae, Osmylidae (Porisminae, Eidoporisminae, Stenosmylinae, Kempyninae: the latter two are Gondwanan), Chrysopidae (Nothochrysinae), Psychopsidae, Myrmeleontidae (Stilbopteryginae), Nemopteridae.

The Oriental/eastern Palaearctic transitions to Malesia are less well defined, largely reflecting paucity of documentation. Palparine antlions (Palparini, tribe of Myrmeleontidae) are predominantly African (Hölzel, 1972), and any Malesian taxa of this group are on the very fringe of the group's range; they are more strongly represented on the northern Indian subcontinent and to the north of Malesia. This form of 'range edge representation' appears to be the case also for some Ascalaphidae, although (as with Myrmeleontidae such as *Hagenomyia*) particular genera occur throughout the region from Malaysia to Australia and are represented by different species in different parts of this wide range. Widespread Palaearctic or Holarctic genera also occur, with a few (such as *Myrmeleon*) being almost cosmopolitan.

Some possible parallels exist between distribution of Neuroptera and the puta-

TABLE 4

The species richness of families of Neuroptera in the described Australian fauna, endemicity of the Australian fauna, and its overlap with the Malesian fauna (Australian faunal data from New, 1996, 1997).

	TOTAL AUSTRALIAN SPECIES	AUSTRALIAN ENDEMICS	AUSTRALIAN SPECIES ALSO IN MALESIA
Rapismatidae	0	0	0
Coniopterygidae	42	40	0
Sisyridae	6	6	0
Osmylidae	31	30	0
Mantispidae	47	45	2
Berothidae	25	25	0
Dilaridae	0	0	0
Hemerobiidae	34	25	4
Chrysopidae	60	53	3
Nymphidae	21	20	?1
Ascalaphidae	37	35	2
Myrmeleontidae	211	206	5*
NOT IN MALESIA			
Ithonidae	13	13	0
Nevrorthidae	1	1	0
Nemopteridae	8	8	0
Psychopsidae	13	13	0

* or 6 if Christmas Island (Australia) is regarded as Malesia.

tive botanical colonisation of Central Malesia. Van Steenis (1972) noted three major 'tracks' for flora (fig. 14) as (a) southward from the Philippines, (b) eastward from Malaysia through southern Indonesia and (c) westward from Melanesia through New Guinea. All such divisions and relationships can be postulated for elements of the Neuroptera as reflected in their currently known distributions. The centres of origin of some lacewing groups (such as the various poorly defined genera of Mantispidae shared between Australia and Malesia) are not yet clear. Current distributions of some taxa, such as Psychopsidae, are anomalous in their presence in areas surrounding or opposing Malesia and absence within Malesia. Thus, Psychopsidae occur to the north (Myanmar, Laos, southern China, Taiwan) and south (Australia). Despite the lack of any species being known from any part of Malesia, Oswald (1993b) suggested that Southeast Asian psychopsids are most likely to have reached mainland Asia by dispersal from Australia through the archipelago. The distribution of Nevrorthidae is also anomalous, and much synthesis remains to clarify the origins and geographical affinities of many Neuroptera in the region.

The peculiarities of the region include Rapismatidae (extending to the Himalayas) and (apparently) the following genera: *Paraconis* (Coniopterygidae), *Thaumatosmylus* (Osmylidae), *Joguina, Nobilinus, Synthochrysa, Borniochrysa, Kostka* (Chrysopidae), *Umbranymphes* (Nymphidae), *Protacheron* (Ascalaphidae), *Episalus* (Myrmeleontidae). The geographical affinities of other, more anomalous, taxa can be clarified only when further material is available for appraisal. Some of those shared with Taiwan are apparently Palaearctic incursions, which occur in Malesia mainly at higher altitudes.

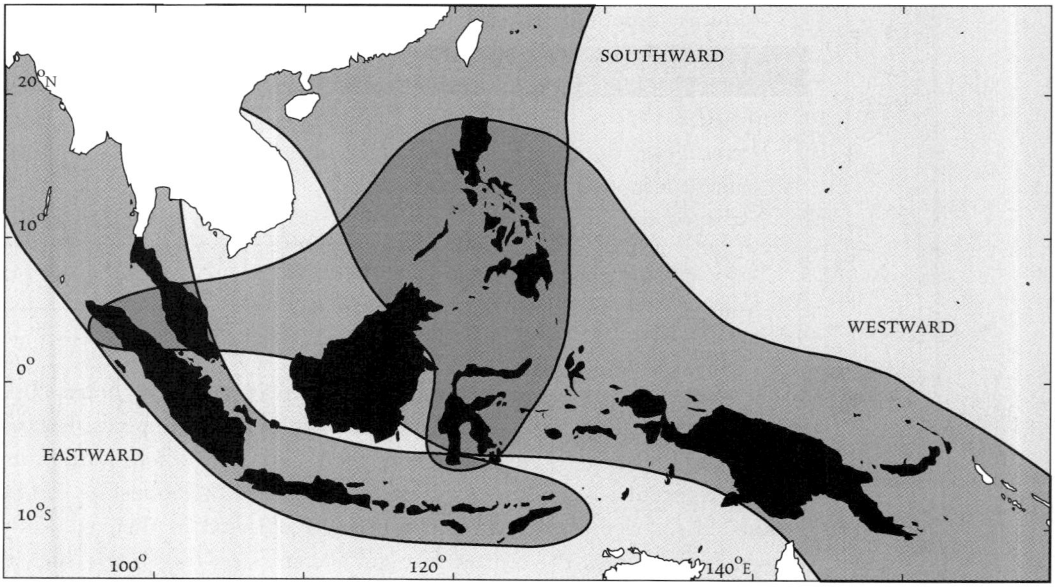

FIGURE 14
Possible tracks for origins of Malesian floral elements (after Van Steenis, 1972).

chapter 4

The families of Neuroptera

For each family of Neuroptera known from Malesia a formal diagnosis is followed by general comments on knowledge on diversity and distribution. Keys to subfamilies and genera, with brief diagnoses of these, are provided, except for the monogeneric Rapismatidae, and for Berothidae due to systematic difficulties. Comments on each genus, with taxonomic uncertainties indicated, are followed by a brief biological outline of the family. Illustrations for each family are selected to indicate the variety of forms and the suite of characters used in identification and recognition. Selected key major references to the global literature are noted. Technical terms are include in the glossary, and the recorded species of each group are listed in the checklist.

Rapismatidae (no established vernacular name)

Family diagnosis

Medium to large lacewings (forewing length 19-35 mm), sturdy, thick-bodied. Head partially retracted beneath prothorax; eyes large, globular; ocelli absent; antennae short, rarely more than about 10 mm, moniliform or slightly serrate; mandibles short and broad; palpi short. Pronotum broad and shield-like. Legs short, lacking spines and spurs; tarsal claws broad basally, with trace of subapical tooth. Wings long and broad, subequal, with small trichosors, usually pale yellowish or green, membrane sometimes mottled. Forewing: costal cell broad, numerous costal crossveins (commonly forked or interlinked) and a recurrent humeral crossvein; Sc and R1 parallel, linked by many crossveins, run free to wing margin; Rs + MA separates from R1 close to wing base; Rs pectinate; MA multiply forked, separates from Rs near base; MP forked near base, with MP2 usually multiply forked; CuA and CuP forked; three anal veins and a simple jugal vein. Hindwing rather similar to forewing, but costal crossveins usually less extensively forked and interconnected, costal area relatively narrow. Both wings with single nygma between base of Rs + MA and MP. Abdomen broad, well sclerotised; cercal callus small.

Female: subgenital plate broad, with median notch; spermatheca simple; lateral gonapophyses rounded.

Male: gonarcus well developed; gonocoxites attached at sides of gonarcus, sometimes bilobed; lobes may be ornamented; mediuncus lobes fused, prominent; hypandrium internum small.

FIGURES 15 - 18

Rapismatidae - *Rapisma*: **15** forewing of *R. barnardi* — **16** female subgenital plate, ventral aspect, of *R. sahbanum* without median projection in apical excavation, and **17** of *R. barnardi* with median projection in apical excavation — **18** male genitalic complex, exemplified by *R. malayanum*, posteroventral aspect (after Barnard & New, 1985; New, 1985e).

15

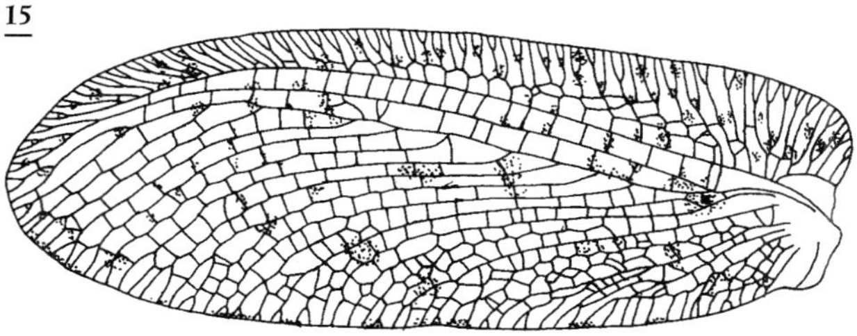

17

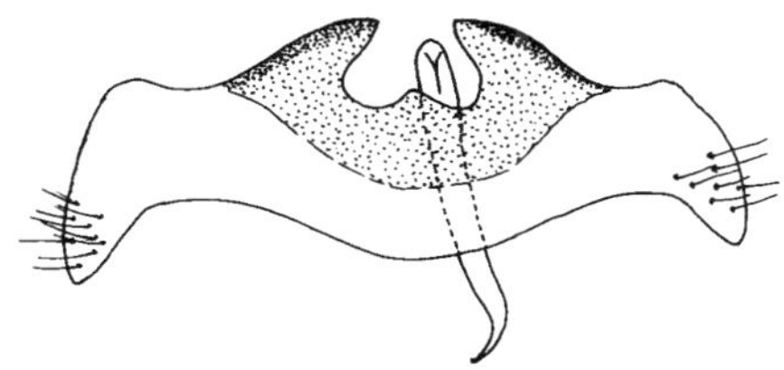

16

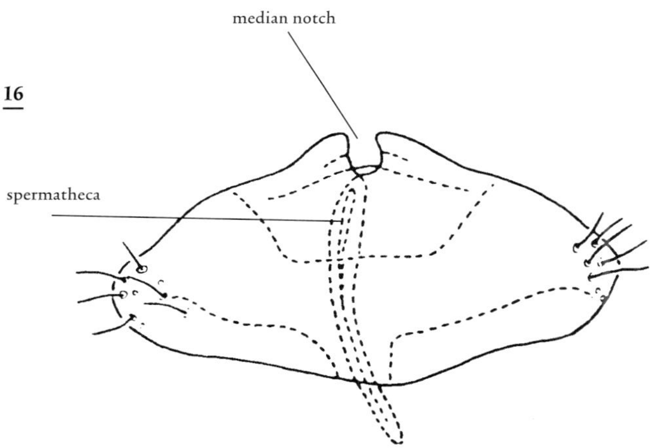

18

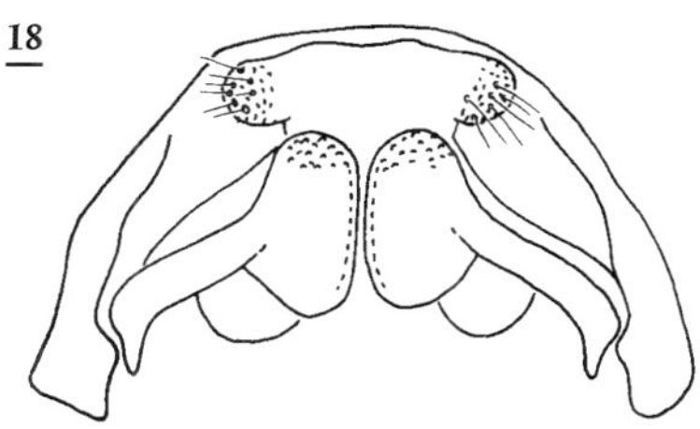

Diversity and distribution

Rapismatidae includes fifteen described species in the world, most of them known from very few specimens (with their combined total of around 35 specimens known) and single localities. The family range extends from the Himalayas (India, Nepal) to parts of Thailand, Malaysia and Indonesia. Nine described species of *Rapisma* have been found in Malesia. In the adjacent part of northern Thailand another one is found, and also one in Myanmar. All may be expected to occur in Malesia. Some species have been found only at higher altitudes. Initial confusion over the type locality of the type species of the sole genus (*Hemerobius viridipennis* Walker), given as 'East Indies' as a misinterpretation of the label 'E. Ind.' abbreviated from 'East India', was clarified by Barnard (1981).

The Rapismatidae are among the most complex and unusual lacewings in the region, and are generally very rare insects. They are among the most characteristic lacewing families in Malesia and the only one predominantly restricted to it. The discovery and study of rapismatid larvae is one of the greatest challenges remaining in clarifying the understanding of family-level phylogeny in the order. Although suggested to be allied to Ithonidae, this supposed relationship is not accepted universally, and the early history of classification of the family (Barnard, 1981) reveals a succession of separate and joint affiliations. Barnard believed that many of the supposed similarities between the families were symplesiomorphies and, following that appraisal, they are treated separately here.

Notes on the genus

Rapisma McLachlan (type species *Hemerobius viridipennis* Walker). All known species of the family are referred to *Rapisma*, and are unlikely to be confused with any other Neuroptera in the region due to their squat, short-bodied appearance, broad head retracted under the prothorax, and broad wings with complex, dense venation (fig. 15). Most are green or brown in life, though the green fades in many dried specimens. Recognition can be hampered by sexual dimorphism in wing shape and head size in the few species for which both sexes are known. The species are separated on features of genitalia.

One of the Malaysian forms, *R. barnardi* New, is relatively distinct; *R. sahbanum* Barnard (fig. 16) is probably closely related to this species. Six others are closely linked and usually referred to as the '*R. malayanum*-complex' (Barnard & New, 1985; figs. 17-18). Malaysia at present appears to be the major evolutionary centre for the family, with the greatest number of species from the highland regions of West Malaysia, and more information on that fauna is a priority for understanding the relationships of this puzzling family.

Biological information

Rapismatidae is the only family of Neuroptera for which early stages and biology are completely unknown.

MAJOR REFERENCE — Barnard (1981): revision and review of relationships.

Coniopterygidae (duskywings)

19

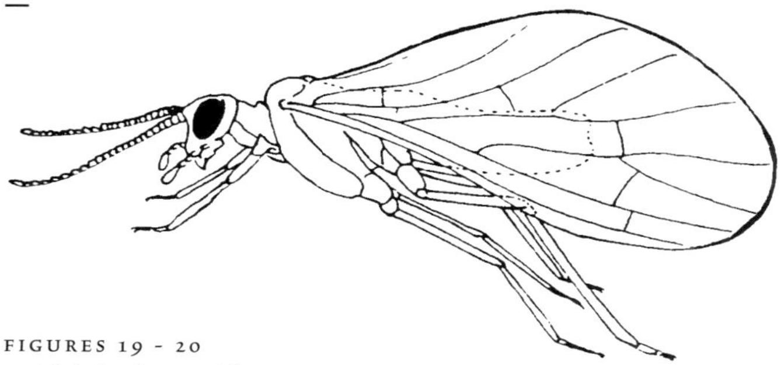

20

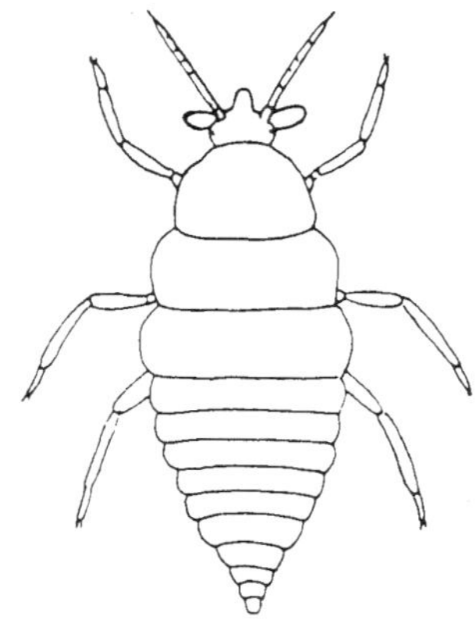

FIGURES 19 - 20
19 Adult Coniopterygidae
20 larval Coniopterygidae.

Family diagnosis

Small lacewings (forewing length usually approximately 2-4 mm, rarely up to about 5 mm), with wing venation greatly reduced; body, wings and legs often covered with grey or white wax. Head with temporal suture well developed; clypeus and labrum short; genae small (Coniopteryginae) or larger (Aleuropteryginae); frontal area between antennae sometimes unsclerotised; ocelli absent. Antennae usually moniliform to filiform, flagellar segments sometimes discoidal; with c. 16-60 segments, the number sometimes diagnostic; scape and pedicel in males sometimes ornamented. Mandibles small; maxillae well developed; lacinia with comb of long hairs along inner edge; galea one-segmented (Coniopteryginae) or three-segmented (Aleuropteryginae); maxillary palpi five-segmented. Labium well developed; labial palpi three-segmented. Prothorax short, lightly sclerotised; pterothorax more heavily sclerotised. Legs slender, not elaborated. Most species with two pairs of broad subequal wings, hindwings rarely reduced (*Conwentzia*); trichosors absent; membrane pale or variously spotted with brown, grey or black; marginal setae often present, as with macrotrichia on longitudinal veins. Venation: never more than two (basal) costal crossveins. Forewing: Sc forked, with posterior branch resembling distal part of R1; Rs and M fused basally; Rs separates from R1 near middle of wing, usually forked with posterior branch sometimes resembling distal part of M; M usually forked; one or two R-M crossveins; sometimes (Aleuropteryginae) two long setae on M in central region of wing; Cu usually forks

FIGURES 21 - 22
Coniopterygidae, wing venation of subfamilies: fore- and hindwings of **21** Coniopteryginae, **22** Aleuropteryginae.

21

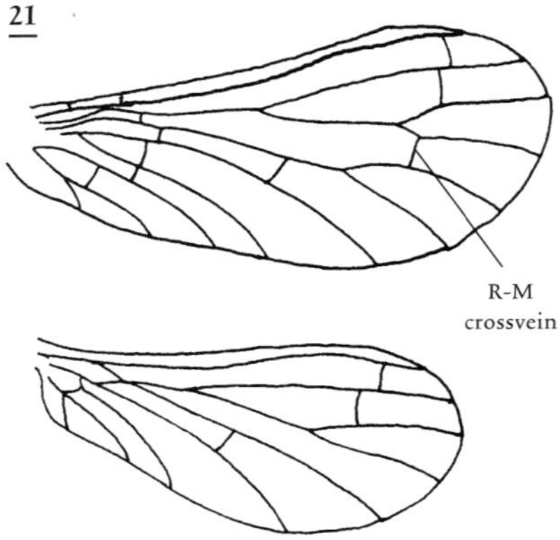

22

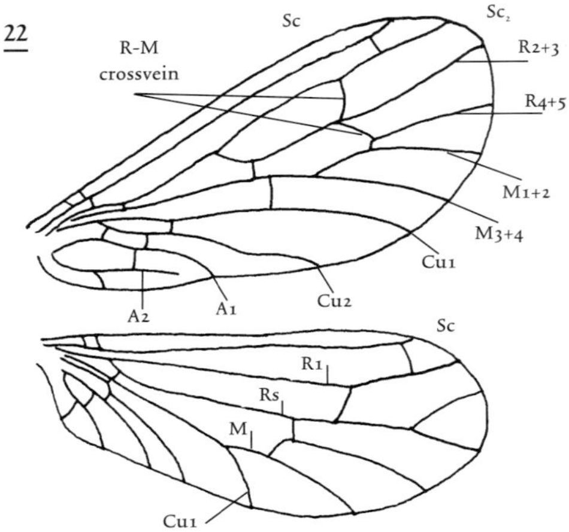

near base, sometimes more distally; one M-Cu crossvein; two anal veins. Hindwing with crossvein from R1-Rs; M forked or simple; R-M crossvein present; Cu forked near base; two anal veins. Abdomen moderately to well sclerotised, except genitalia; wax glands on tergites and sternites; plicaturae present in Aleuropteryginae.
Female: genitalia weakly sclerotised; sternite ix only rarely distinct; lateral gonapophyses short, without styli; subanale (sternite x) often present.
Male: tergite ix and sternite ix fused into sclerotised ring; usually incorporating hypandrium (fused coxopodites ix); styli usually present; gonarcus commonly not distinct; sternite x usually present; parameres elongate; true penis present. No cercal callus or trichobothria.

Coniopterygidae are thus characterised by general appearance, reduced wing venation with no more than two crossveins and no 'end-twigging' to the main veins (figs. 21, 22). An apparently primitive venational feature, shared with Megaloptera and unique within Neuroptera, is that the subcosta forks into two branches; the second branch (Sc2) is reduced in all other lacewings. Adults have no trace of cercal calli, and the complex male genitalia are also characteristic and difficult to homologise with those of other lacewings. In contrast, female genitalia are weakly developed, and their membranous parts sometimes very difficult to differentiate and homologise.

Two distinctive subfamilies occur, and both are well represented in Malesia. The South American genus *Brucheiser* Navás may constitute a third subfamily, but its status can be confirmed only once the larva is discovered. The subfamilies are separable on features of wing venation.

Diversity and distribution

The Coniopterygidae, a family of over 400 described species in the world, with 59 in Malesia, derive their common name from the coating of white or greyish wax on the wings and body, which results in them having an appearance similar to that of whiteflies (Aleyrodoidea). They are by far the smallest Neuroptera, with most species being only about 2-4 mm in body length; Meinander (1972) aptly termed them 'the pygmies of the Planipennia'. The characteristic wax is produced by hypodermal wax glands, unique within the order.

Both subfamilies are widely distributed, but only a restricted set of the genera in each are found in Malesia. The predominant genus in the region is *Heteroconis*; most other genera have been reported from only few species.

Key to the subfamilies of Coniopterygidae

1 Forewings with one R-M crossvein in central region (fig. 21); hindwing vein Rs not branching from R very near base of wing ‣ **Coniopteryginae**

– Forewings with two R-M crossveins in central region (fig. 22); hindwing vein Rs branching from R very near base of wing ‣ **Aleuropteryginae**

CONIOPTERYGINAE

Key to regional genera of Coniopteryginae (after Meinander, 1972)

1 Hindwing vein M unforked (fig. 23) ▸ ***Coniopteryx*** (4)
– Hindwing vein M forked ▸ 2

2 Crossvein M-Cu1 in both wings oblique, usually striking posterior branch of M (fig. 24) ▸ ***Semidalis***
– Crossvein M-Cu1, at least in forewings, striking longitudinal veins at right angles, striking stem of M (fig. 25) ▸ 3

3 Hindwing vein Rs forking from R basal to middle of wings (fig. 25) ▸ ***Neosemidalis***
– Hindwing vein Rs forking from R in middle of wings (fig. 26) ▸ ***Conwentzia***

4 (Males only keyed (fig. 27); all subgenera included, to facilitate recognition.) Gonarcus and hypandrium synscleritous ▸ *Coniopteryx* subg. ***Scotoconiopteryx*** (New World only)
– Gonarcus and hypandrium discleritous ▸ 5

5 Parameres shorter than penis; basal part of penis forming circular structure with styli ▸ *Coniopteryx* subg. ***Metaconiopteryx*** (Palaearctic)
– Parameres longer than penis; basal part of penis not forming circular structure ▸ 6

6 Base of styli attached to gonarcus before apex; ejaculatory duct not sclerotised ▸ *Coniopteryx* subg. ***Xeroconiopteryx***
– Base of styli attached to gonarcus at apex; if not, ejaculatory duct sclerotised ▸ 7

7 Gonarcus divided into two lateral plates ▸ *Coniopteryx* subg. ***Coniopteryx***
– Gonarcus not divided into two lateral plates ▸ 8

8 Styli fused below parameres ▸ *Coniopteryx* subg. ***Protoconiopteryx*** (Australia only)
– Styli not fused below parameres ▸ *Coniopteryx* subg. ***Holoconiopteryx*** (Europe, Africa)

FIGURES 23 - 26

Coniopterygidae - Coniopteryginae: fore- and hindwings of representatives of **23** *Coniopteryx*, **24** *Semidalis*, **25** *Neosemidalis*, **26** *Conwentzia*, with reduced hindwing.

23

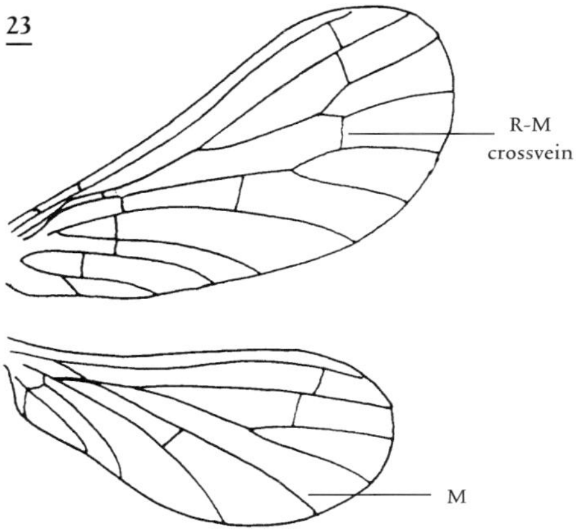

24

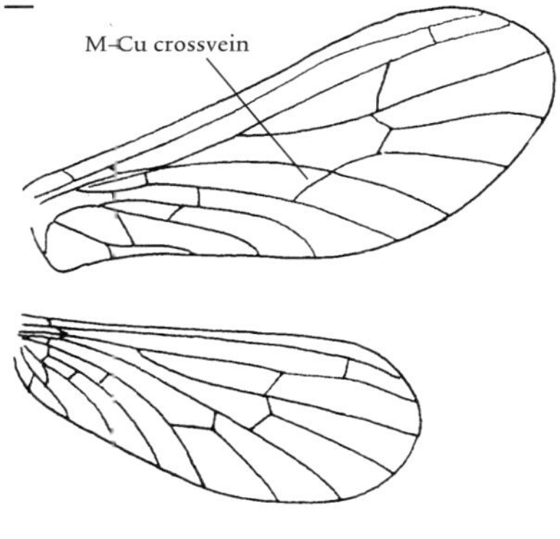

25

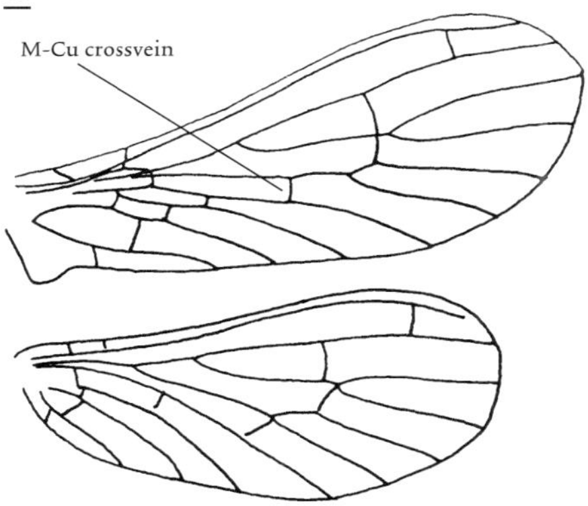

26

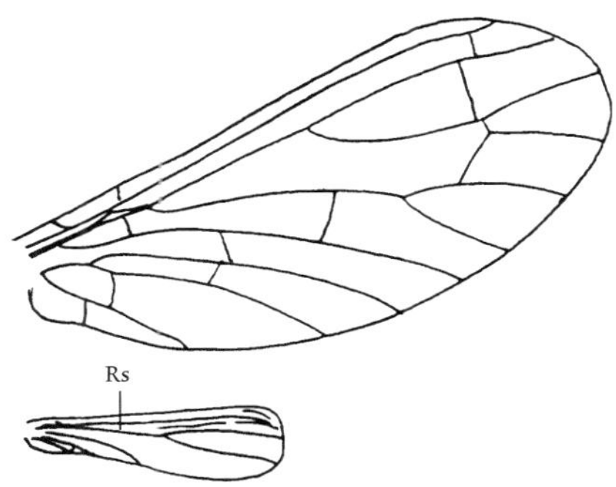

FIGURES 27 - 28

Coniopterygidae - Coniopteryginae: representative apices to show genitalic features, lateral aspect of **27** male and **28** female *Coniopteryx*.

27

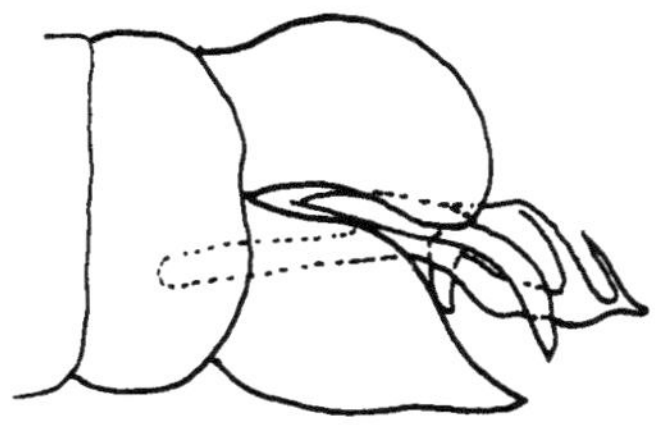

28

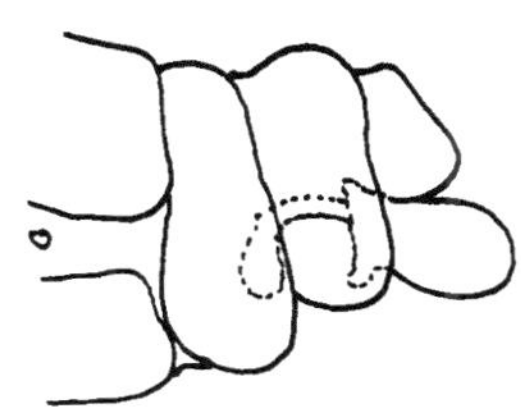

Notes on the genera of Coniopteryginae

Coniopteryx Curtis (tribe Coniopterygini; type species *Coniopteryx tineiformis* Curtis) is a vast genus of worldwide distribution, and most of the 150 or so described species have been allocated to one of a series of subgenera on features of male genitalia. Some species, including several from Malesia, can not yet be allocated confidently, as they are known only from females (fig. 28), but it seems that only two subgenera are represented in the region.
Both subgenera *Xeroconiopteryx* and *Coniopteryx* are widespread. *Xeroconiopteryx* is diverse in Australia and a species from Papua New Guinea (*C.* (*X.*) *papuensis* Meinander) is probably related to Australian taxa. *Coniopteryx* (*Coniopteryx*) is apparently more diverse, with about a dozen species reported from various parts of Malesia, some apparently being widespread. It is likely that many additional species of these poorly known insects will be found in the region.

Conwentzia Enderlein (tribe Conwentziini; type species *Conwentzia pineticola* Enderlein) is a widespread genus, with one species (*C. nietoi* Monserrat) recorded from Indonesia and Sri Lanka. The hindwing of many members of the genus is reduced (fig. 26), and *Conwentzia* is the only genus of the tribe to show this feature.

Neosemidalis Enderlein (tribe Conwentziini; type species *Parasemidalis metallica* Enderlein) comprises the subgenera *Neosemidalis* and *Leucosemidalis* Meinander. Both are predominantly Australian but two species of the former have been described from Papua New Guinea. The two subgenera are separable on the form of the wing covering. In *Leucosemidalis*, the wings are hyaline and covered thickly with white wax and in *Neosemidalis* s.str. the greyish wings are covered sparsely with grains of wax (Meinander, 1972). *Leucosemidalis* has not yet been found in Malesia.

Semidalis Enderlein (tribe Conwentziini; type species *Coniopteryx aleyrodiformis* Stephens) can be divided into five species-groups on male genitalic features. One of these, the '*S. decipiens* group', comprises a small number of East Asian and East Palaearctic species in which the male genitalia have a distinct stylus. The predominant species, *S. decipiens* (Roepke), has been found in West and East Malaysia and Indonesia. It was described originally as *Parasemidalis* Roepke, which is now considered a synonym of *Semidalis* (Meinander, 1972).

FIGURES 29 - 33

Coniopterygidae - Aleuropteryginae: fore- and hindwings of **29** *Heteroconis*, **30** *Coniocompsa*, **31** *Spilococonis*, **32** *Cryptoscenea*, **33** *Paraconis*.

29

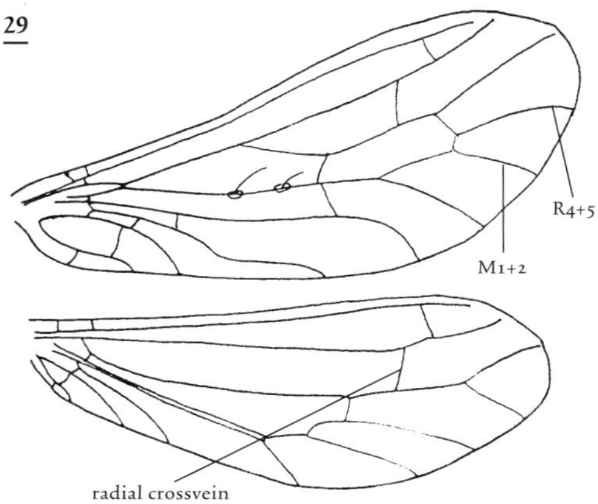

30

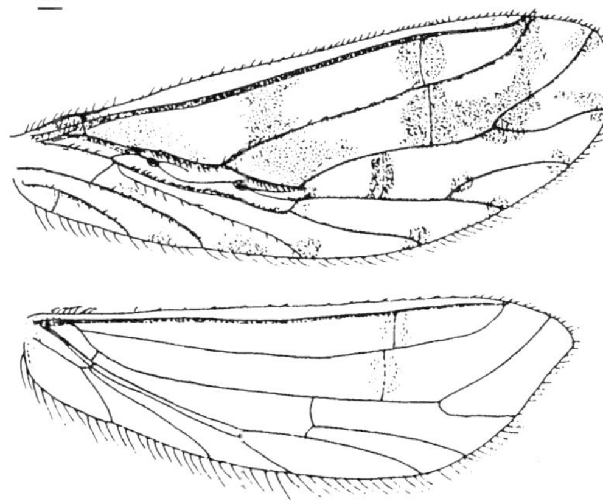

31

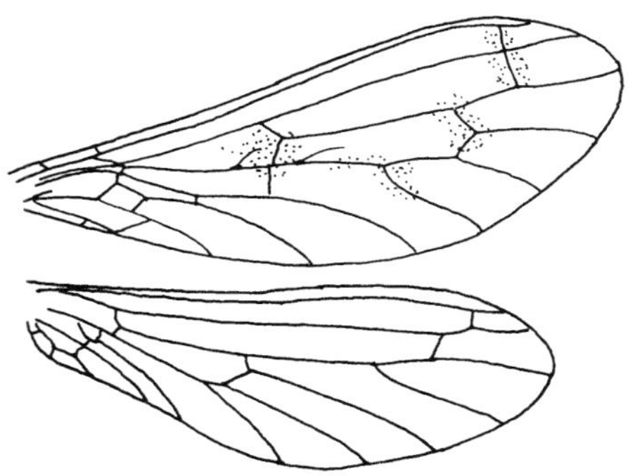

32

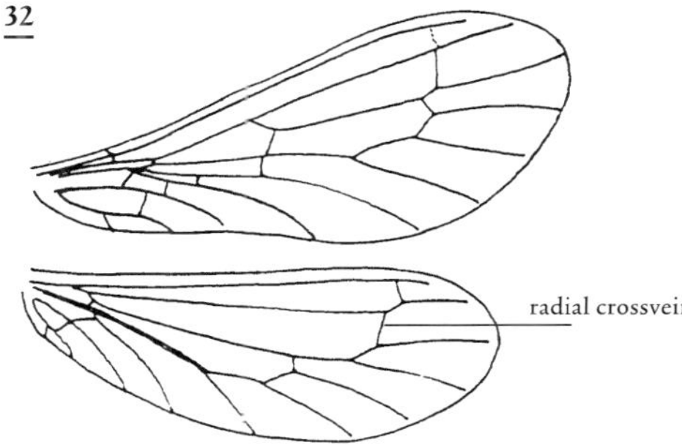

33

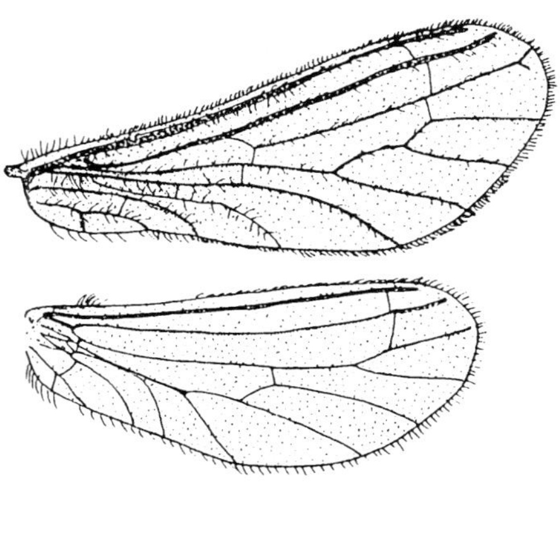

Key to regional genera of Aleuropteryginae (after Meinander, 1972)

1 Hindwing radial crossvein striking Rs on stem or fork (fig. 22, 23) ▸ 2
– Hindwing radial crossvein striking Rs on branch R2+3 (fig. 24-26) ▸ 3

2 M forked in both wings; anterior branch of M in forewing coalescing with vein R4+5 or connected to it by crossvein (fig. 22) ▸ *Heteroconis*
– M not forked in either wing; or forked but anterior branch of M in forewing not connected to R4+5 (fig. 23) ▸ *Coniocompsa*

3 Antennal scape long, at least three times as long as broad ▸ *Spiloconis*
– Antennal scape only slightly longer than broad ▸ 4

4 Veins M and Cu1 in hindwing running very close to each other for more than half their lengths, so no membrane visible between them, then separating at point or distally connected by crossvein. Males only, separable clearly, as follows:

a Eighth sternite of male without plicaturae; ninth segment strongly sclerotised ▸ *Cryptoscenea*
b Eighth sternite of male with plicaturae (fig. 34); ninth segment weakly sclerotised ▸ *Paraconis*

Notes on the genera of Aleuropteryginae

Coniocompsa Enderlein (type species *Coniocompsa vesiculigera* Enderlein) is the only genus in the tribe Coniocompsini, and is very distinctive. In particular the antennal segments have two whorls of long hairs, and the wings are narrow and tapered, with long marginal hair fringes. Vein Sc2 (at least, in the hindwing) curves anteriorly toward the wing apex; the stem of forewing vein Cu is rather long, Cu2 not connected by crossveins to Cu1 or A1; forewing membrane commonly spotted. Both sexes have plicaturae on abdominal segments III-V.

Coniocompsa is widely distributed in the tropics, and several species occur in Malaysia and/or Indonesia. *C. meinanderi* Monserrat is the most widely reported species: Malaysia (East and West Malaysia, described from Penang), Singapore and Indonesia (Krakatau Islands). Both it and *C. traceyae* New have been found in Sabah.

Spiloconis Enderlein (tribe Fontenelleini; type species *Spiloconis sexguttata* Enderlein) is a small genus. The elongate antennal scape is a very characteristic feature. A single record of *S. sexguttata* Enderlein from Thailand (Monserrat, 1982) is the only indication of its possible presence in western Malesia; the genus is also known from the Philippines.

Paraconis Meinander (tribe Fontenelleini; type species *Paraconis borneensis* Meinander) is known from only a single species, described from Sabah. It is very similar to *Cryptoscenea*, but differs in the arrangement of plicaturae (present on abdominal segment VIII).

Cryptoscenea Enderlein (tribe Fontenelleini; type species *Helicoconis australiensis* Enderlein) is predominantly an Australian genus, with six species. One, *C. novaeguineensis* Meinander, occurs in Papua New Guinea, and Monserrat (1982) reported an undescribed individual from Bali.

Heteroconis Enderlein (tribe Aleuropterygini; type species *Heteroconis ornata* Enderlein) is the largest genus in the subfamily. It contains around 60 described species, mostly from Australia and Malesia, with a few representatives in tropical Africa. It is undoubtedly the predominant genus of Coniopterygidae in Malesia. Some species have relatively uniformly tinted wings (fig. 35) but many of the species have characteristically spotted wings (fig. 36) and antennae with groups of segments contrasting black and pale, so that the antennae appear banded. A few species from Papua New Guinea and Indonesia have strongly pointed, almost falcate wings and have been placed in a distinct subgenus, *Drepanoconis* Tjeder (type species *H. (D.) amoena* Tjeder; fig. 37). Meinander (1972) tentatively considered the Australian species to be generically distinct from those in Southeast Asia, but this is clearly not so, as the many additional species now known adequately bridge the extremes of form. Undoubtedly, many further species will be found in Malesia.
Heteroconis is closely related to *Aleuropteryx* Löw (not reported from Malesia, and almost wholly Holarctic), but is separable clearly on wing features and the number of antennal segments (18, rarely 17 in *Heteroconis*; 21-27 in *Aleuropteryx*).

FIGURES 34 - 37

34 Appearance of plicaturae on abdominal sternites of Aleuropteryginae — Coniopterygidae - *Heteroconis*, to indicate variety of appearance in this predominant regional genus: **35** *H. spinosa* New as representative of uniformly coloured taxa, **36** *H. fenestrata* New as representative of strongly patterned taxa, **37** subgenus *Drepanoconis*: *D. amoena* Tjeder, with tapered forewing (after New, 1990c).

34

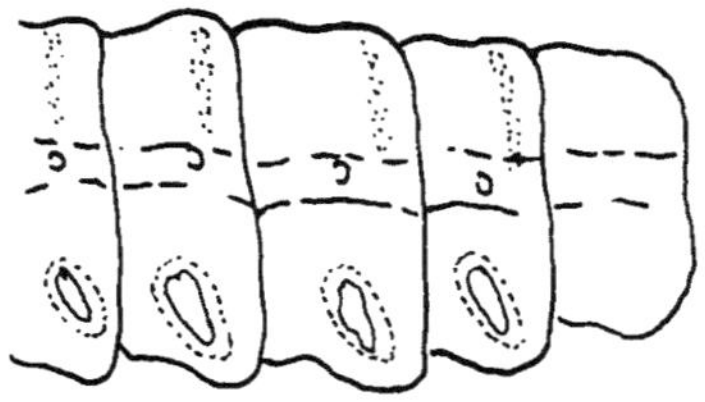

35

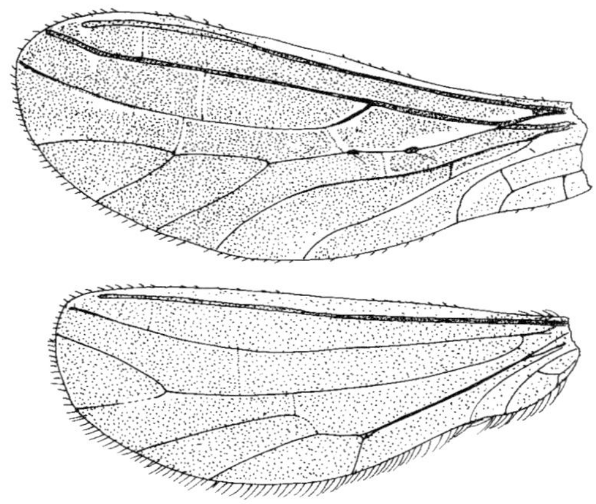

36

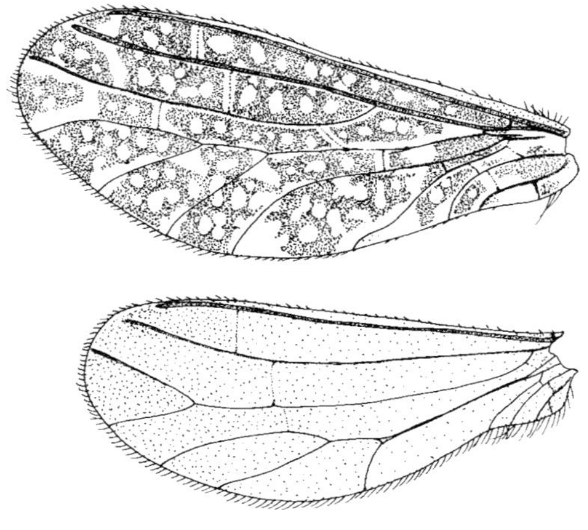

37

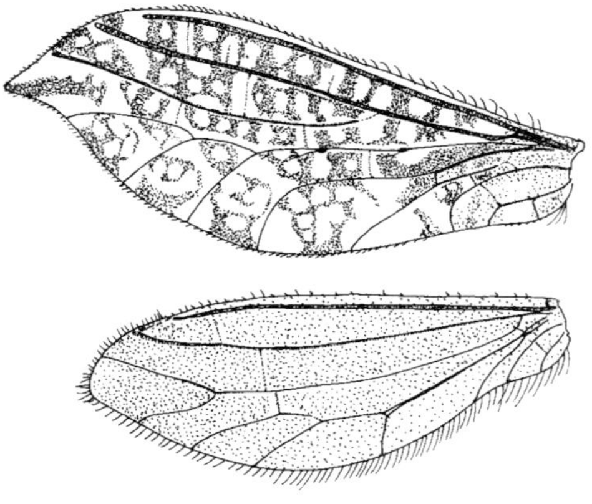

Larval key to the subfamilies of Coniopterygidae

NOTE

Early stages and biology of the region's Coniopterygidae are very poorly known, and there are no formal descriptions of larvae of Malesian species. Nevertheless, the key to larvae of the two subfamilies provided by Meinander (1972) seems likely to be valid, and is given here as an initial guide.

1 Larvae with jaws not projecting from beneath labrum; antennae about twice as long as labial palpi ▸ **Coniopteryginae**

2 Larvae with jaws longer, projecting from beneath labrum; antennae about as long as labial palpi ▸ **Aleuropteryginae**

This key is based almost entirely on species from the Holarctic region, but incorporates *Cryptoscenea*, among others. Larvae of *Spiloconis*, *Paraconis* and *Heteroconis* have not been described, but the general form of larvae of taxa from elsewhere is likely to facilitate family recognition.

Biological outline

Coniopterygidae occur widely on vegetation, and are captured most frequently by beating and sweeping. In Singapore, Murphy & Lee (1971) captured several species from forest canopy on gum-coated balloons. Larvae occur in association with phytophagous mites and other small, soft-bodied arthropods, but have been reported only infrequently from Malesia.

MAJOR REFERENCES — Meinander (1972): world revision of family, with summary of all information published previously on the family; Meinander (1990): checklist of world species.

Sisyridae (spongeflies)

38

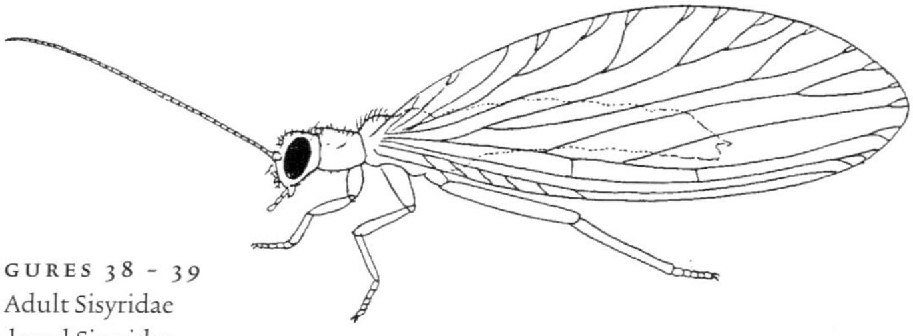

39

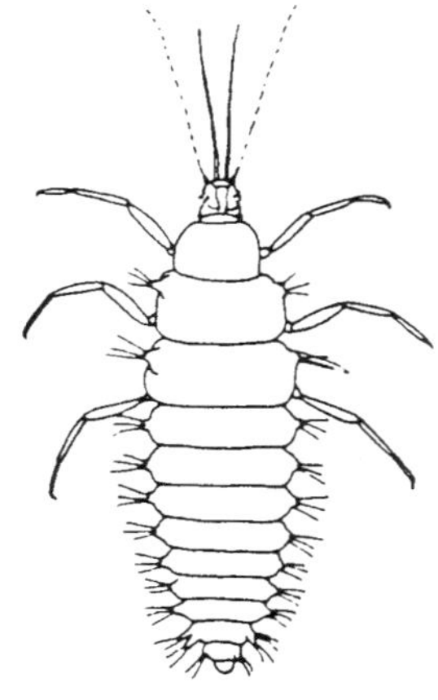

FIGURES 38 - 39
38 Adult Sisyridae
39 larval Sisyridae

Family diagnosis

Delicate lacewings, small to medium-sized (forewing length 4-10 mm). Head short and rounded; ocelli absent; antennae moniliform to filiform, usually about half length of forewing; mandibles acutely pointed, usually asymmetrical; maxillary palpi long, five-segmented; labial palpi three-segmented. Pronotum short, broad; pterothorax large; legs long and slender, claws simple. Wings subequal, oval; trichosors on distal half, sometimes only weakly defined on hindwing; nygmata absent; wings sometimes mottled, usually pale grey or pale brown. Costal space narrow, with few simple crossveins, mainly in basal half of wing; no recurrent humeral crossvein; pterostigmal crossveins obscure; veins and margins with long macrotrichia. Forewing: Sc and R1 indistinctly fused near apex, no basal crossveins between these; Rs and R1 separate near base of wing; apparent Rs with several branches and linked to R1 by two (forewing) or one (hindwing) crossvein(s); posterior branches forked; few, if any, gradate crossveins; most more posterior veins forked. Hindwing CuA long. Abdomen weakly sclerotised, tergites and sternites small; ectoprocts small, with cercal callus of trichobothria.

Female (fig. 40): tergite ix extended laterally; a pair of fused lateral gonapophyses .

Male (fig. 41): pair of elaborate appendages (? entoprocessus); ectoprocts very small; gonarcus, parameres, hypandrium internum present.

FIGURES 40 - 43
Sisyridae - general form of genitalic structures of *Sisyra* of **40** female, **41** male — **42** fore- and hindwing of *Sisyra* — **43** ventral aspect of central abdominal segments of Sisyridae larva to show gills.

40

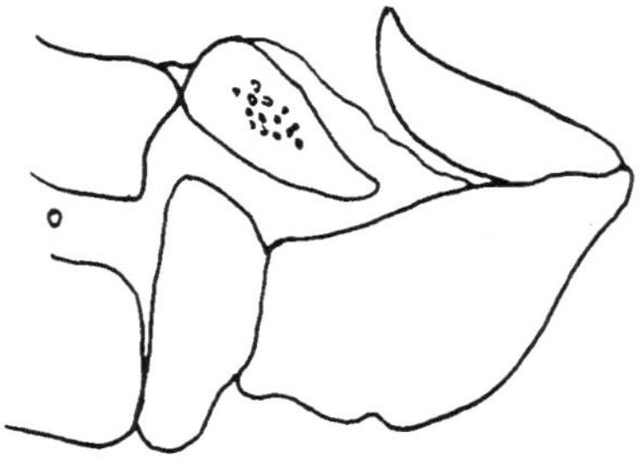

41

42

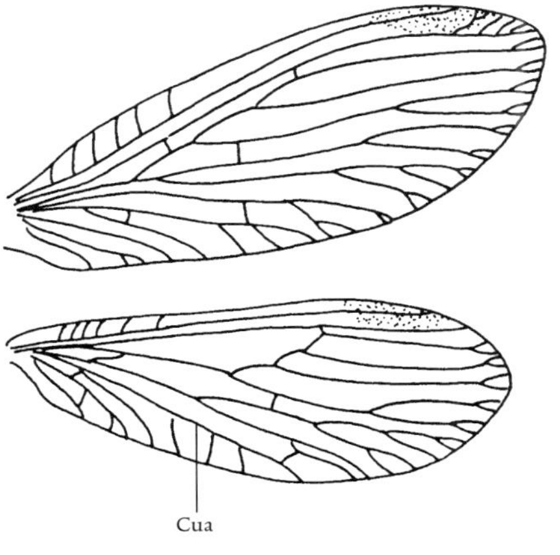

43

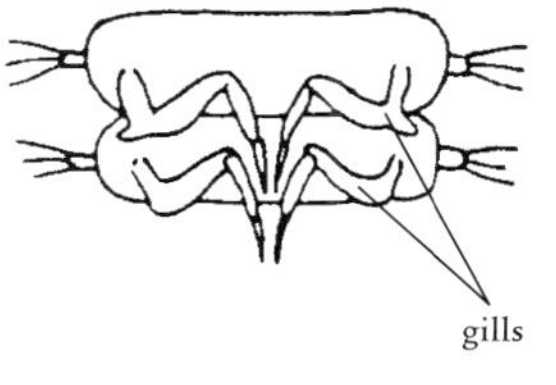

Diversity and distribution

The Sisyridae is one of the smaller families of Neuroptera, with three species in Malesia and fewer than 50 species known in the world. Sisyridae of the Oriental region were reviewed by Monserrat (1981). All records from Malesia are of the cosmopolitan genus *Sisyra* Burmeister. However, a species of *Sisyrina* Banks has been found in southern India, and a second one in northern Australia, so the genus is included in the following key to facilitate recognition.

Key to the genera of Sisyridae

1 Fore- and hindwings with, at most, a single crossvein in apical third (fig. 42) ▸ *Sisyra*
– Fore- and hindwings with a series of crossveins in apical third ▸ *Sisyrina*

Note on the genus

Sisyra Burmeister (type species *Hemerobius fuscatus* Fabricius). Few species of *Sisyra* occur in Malesia. The most widespread species is *S. indica* Needham (Philippines, West Malaysia, southern Thailand), of which *S. vigana* Navás is a synonym. *S. maculata* Monserrat, described from Phuket (Thailand) has not yet been reported elsewhere. *Sisyrella* Banks is a synonym of *Sisyra*.

Biological outline

Sisyridae are invariably associated with freshwater habitats, usually with well-aerated streams and rivers. Eggs are laid on vegetation overhanging water into which the hatchling larvae drop and swim actively in search of a host: freshwater sponges and, possibly, also bryozoans, which tend to thrive only in clean waters. The larvae, which have a characteristic form (fig. 39), are unique among Neuroptera in having ventral abdominal processes (fig. 43). They leave the water to pupate. Adults appear to be rather weak flyers; they are usually found on waterside vegetation rather than further afield. Most species are rather elusive, and identification is not always certain. Unidentified sisyrid larvae have been found in Java (Esben-Petersen, 1933) and this stage is the most frequent evidence of the incidence of the family.

Adult Nevrorthidae (no established vernacular name; larva fig. 9a) are very similar to Sisyridae, but the wings are not hairy and some forewing coastal crossveins are forked. The larvae are very different from larvae of Sisyridae (cf. fig. 39): they have shorter jaws and lack ventral abdominal gills.
There are nine described species of Nevrorthidae in the world. No representatives have been found yet in Malesia, but they are known from southern Europe, Taiwan, Japan and eastern Australia and could thus, possibly, be found in streams in New Guinea.
Adults occur in damp habitats, often near streams. The larvae live on the bottom of fast-flowing streams.

MAJOR REFERENCES — Parfin & Gurney (1956): good synopsis of earlier knowledge; Tjeder (1957): revision of South African fauna, with summary of earlier global literature; Monserrat (1977): world catalogue of family; Pupedis (1987): feeding biology of adults.

Osmylidae (no established vernacular name)

44

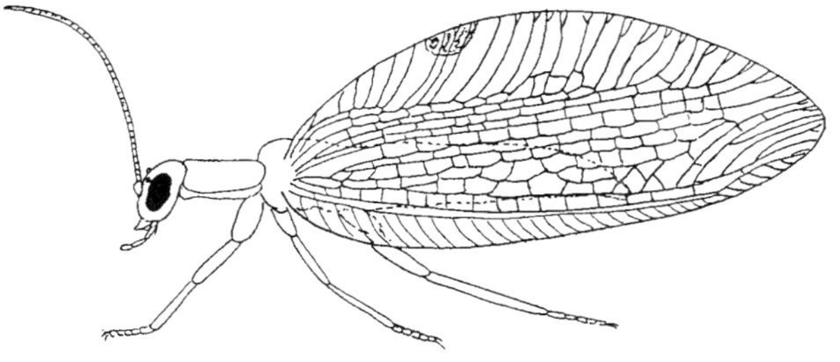

45

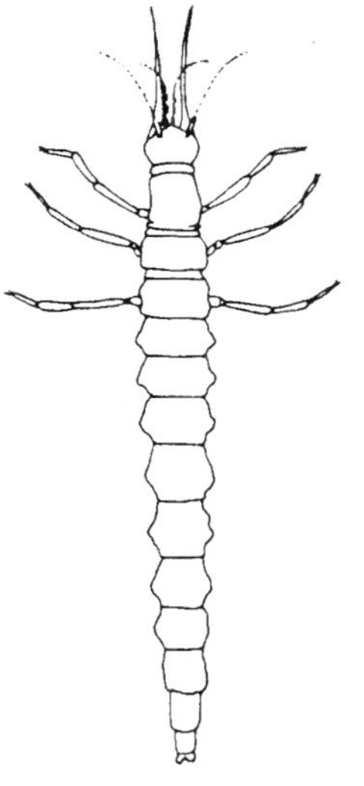

FIGURES 44 - 45
44 Adult Osmylidae
45 larval Osmylidae

Family diagnosis

Slender, medium to large lacewings (forewing length approximately 15-30 mm). Head short; eyes moderately large; three ocelli usually present, at least represented by raised tubercles, sometimes indistinct; antennae filiform, usually about 1/3-1/2 length of forewing; mandibles pointed, usually asymmetrical; maxillary palpi long, five-segmented; apical segment of three-segmented labial palpi with outer sensory field. Pronotum usually narrower than pterothorax. Legs slender, long, claws simple. Wings subequal, elongate, broadly oval or (rarely in regional fauna) slightly falcate; trichosors present except near base; forewing, in particular, may be variously marked and can have one or two embossed spots on posterior margin; usually two nygmata on each wing. Costal space relatively broad, with numerous crossveins (simple, some often branched, sometimes interconnected); forewing humeral crossvein not recurrent; pterostigma distinct; Sc and R1 closely parallel, coalesce behind pterostigma, usually with only basal Sc-R1 crossvein, but cell often with dark spots or streaks; Rs and MA with common stem; Rs pectinate; MP forked; Cu forked near base of wing; three anal veins present; whole wing with numerous crossveins, mostly not forming distinct gradate series. Abdomen with well defined tergites and sternites; cercal callus with numerous small trichobothria.

Female (figs. 46, 47): tergite ix prolonged ventrally; lateral gonapophyses long and slender, with small apical style; spermathecae present, often with long coiled duct.

Male (fig. 48): ectoprocts usually fused dorsally; gonarcus complete, usually with entoprocessus; parameres free or fused; sometimes an arched subarcus above parameres; small hypandrium internum present.

The subfamilies of Osmylidae are separated mainly on wing venation. Spilosmylinae are diagnosed by: forewing costal crossveins simple, occasionally with a few forked (Osmylinae and Kempyninae have many forewing costal crossveins forked); forewing with a single basal subcostal crossvein, occasionally with traces of a few more (Porisminae, from Australia, have numerous subcostal crossveins); forewing vein MP forking near base of wing (unlike Stenosmylinae and Eidoporisminae, in which this vein forks at or beyond half length of wing); numerous crossveins, usually at least six series basal to two main outer gradate series (Protosmylinae have far fewer such crossveins basal to the main gradate series).

Spilosmylinae is by far the largest subfamily of Osmylidae, and many of the other subfamilies noted above contain very few species. However, generic separations in the group are sometimes ambiguous. Many of the genera recognised on small venational features by Krüger (1913) were synonymised with *Spilosmylus* Kolbe by Kimmins (1942), a step endorsed by Tjeder (1957) in his treatment of the limited southern African fauna. *Spilosmylus* is thus both diverse and complex. It includes a variety of species, ranging from rather delicate pale green forms to more robust brown species. Most of the described species of Osmylidae from Malesia are known from few specimens and, in many cases, from single localities. *Thaumatosmylus* Krüger is very similar to *Spilosmylus*, but was not synonymised by Kimmins or Tjeder, and was accepted as valid by New (1991a), albeit with some 'blurring' of characters used earlier to separate the genera; some species are hard to allocate to one or other genus, but the whole of the Malesian fauna seems to fall firmly into this complex.

Diversity and distribution

Globally, Osmylidae are represented by several distinctive subfamilies. The Osmylidae of Malesia are characterised by the absence of most groups and the major radiation of one subfamily, Spilosmylinae, which contains many closely related species particularly in Indonesia and Papua New Guinea (New, 1986b, 1988b, 1991a). The region is clearly a major centre of evolution for this group. The southern subfamilies of Osmylidae found in Australia appear to be entirely absent; the Osmylinae (which extend in the Palaearctic to Japan and Taiwan) are likewise not represented.

FIGURES 46 - 49

Osmylidae - Spilosmylinae: representative genitalic structures, abdominal apex, lateral aspect of **46** female *Spilosmylus sibilensis* New, **47** female *S. sinuosus* New, and **48** male *S. sibilensis* — **49** fore- and hindwing of *S. sinuosus* (from New, 1988b).

46

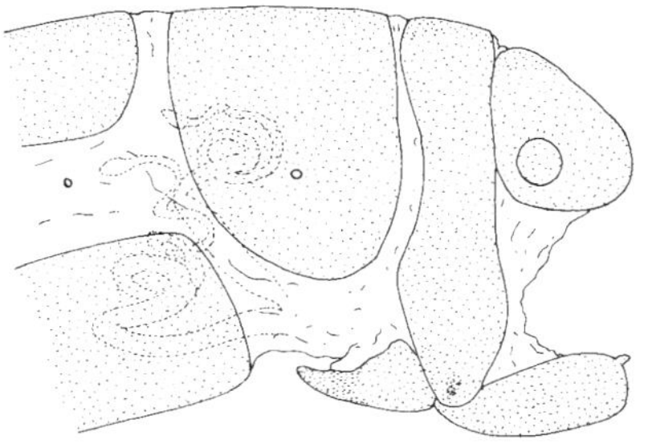

47

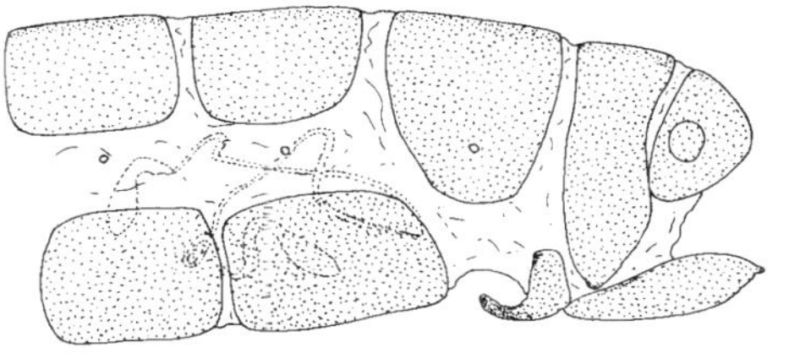

48

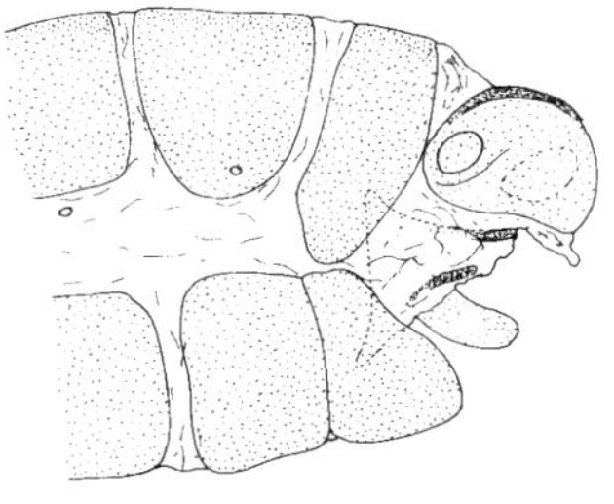

49

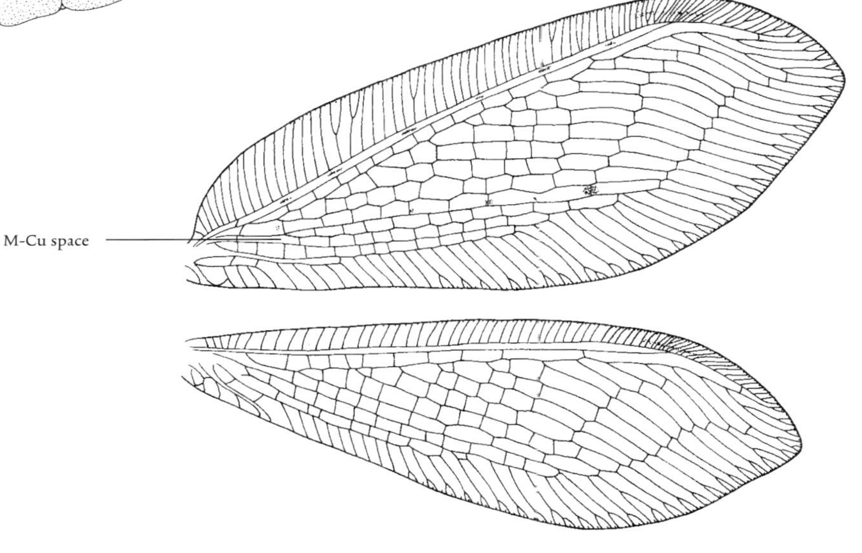

Key to the genera of Osmylidae

1 A clear space free of crossveins following basal M-Cu crossvein in forewings; (forewings sometimes with embossed patches; figs. 49-52) ▸ ***Spilosmylus***

– No clear space free of crossveins beyond basal M-Cu crossvein in forewings; (forewings never with embossed patches; fig. 53) ▸ ***Thaumatosmylus***

NOTE

As indicated by New (1991a), the crossvein character is not wholly reliable. Generally, the wings of *Thaumatosmylus* are rather broad, usually with brownish or greyish markings, and the hindwing is usually very broad centrally and tapers to a distinct acute apex, unlike the usual narrower, straplike form in *Spilosmylus* (cf. figs. 49 and 53).

Notes on the genera

Spilosmylus Kolbe (type species *Osmylus tuberculatus* Walker) is by far the most diverse genus of Osmylidae (and, indeed, of Neuroptera, as currently documented) in the region. It has undergone considerable radiation in Malaysia, Indonesia, Papua New Guinea and the Philippines, but only two species occur in Australia, both confined to the tropical north of the country. As noted above, several apparently distinct 'forms' are included in the genus. The small pale green taxa (some formerly placed in *Conchylosmylus* Krüger) often have a conspicuous raised/embossed spot (fig. 50) on the posterior margin of the forewing; its function is unknown, but it is present in both sexes. Many of these species are associated with forests, sometimes at higher altitudes. Larger brown taxa also occur, some of them with gold iridescence on the wings (as in *Glenochrysa*, Chrysopidae). Their wings tend to be relatively narrower than in the green taxa, and their bodies more robust. Many species are characterised by intermittent darkening of forewing veins Sc and R1, sometimes accompanied by parallel darkening of the membrane in cell Sc.
Regional generic synonyms include *Rhipidosmylus* Krüger, *Conchylosmylus* Krüger, *Stigmatosmylus* Krüger, *Grammosmylus* Krüger, *Heliosmylus* Krüger and *Centrolysmus* Navás.

Thaumatosmylus Krüger (type species *Osmylus diaphanus* Gerstaecker) contains about five described species, most of them apparently rare. Other than one species from Taiwan, *Thaumatosmylus* occurs only in Malaysia and western Indonesia and, apparently, does not extend to New Guinea.

FIGURES 50 - 53

Osmylidae - Spilosmylinae: wings of representative species of *Spilosmylus* to indicate diversity of **50** *Silosmylus sinuosus* New, with embossed spot on forewing, **51** *S. javensis* New, **52** *S. conspersus* (Walker), **53** *Thaumatosmylus*; NOTE triangular hindwing of *Thaumatosmylus* broader compared with *Spilosmylus*.

50

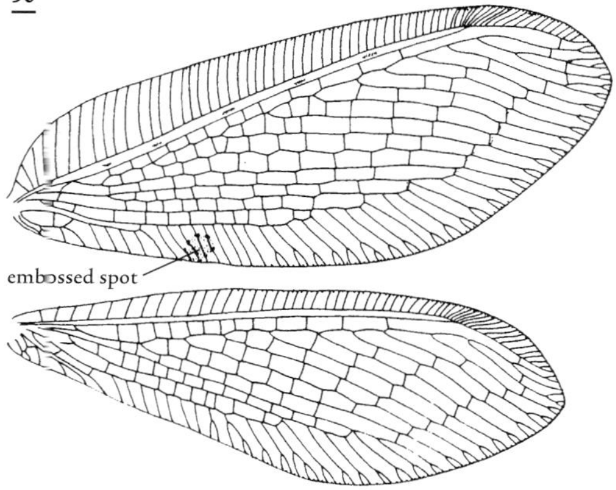

51

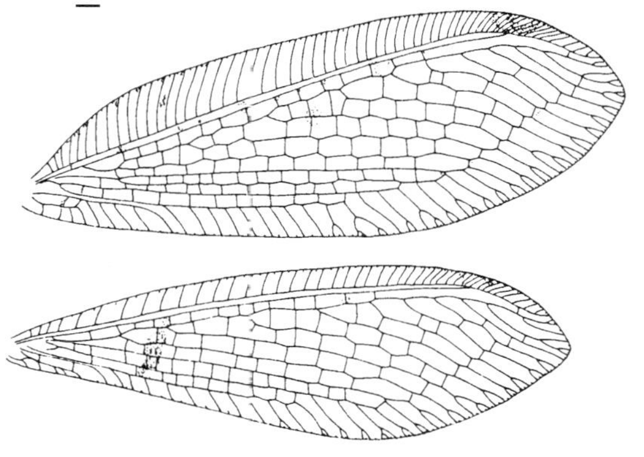

52

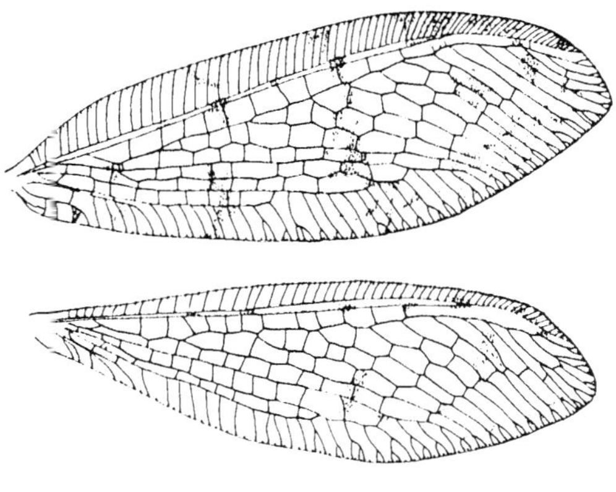

53

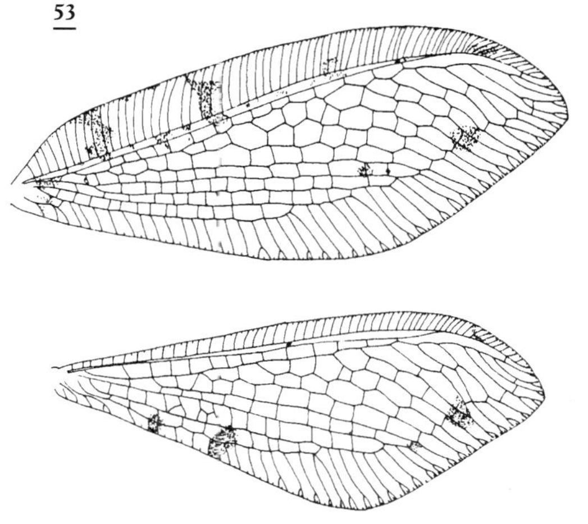

Biological outline

Osmylidae are generally rather scarce insects and are usually collected at lights or by beating or sweeping vegetation. About 150 species have been described worldwide, with 56 in Malesia. Virtually nothing has been reported on the biology of Spilosmylinae, and it is presumed that their larvae may be arboreal, perhaps resembling those of some Australian Osmylidae in occurring on and under bark as predators of arthropods. In contrast, Osmylinae are sometimes more closely associated with water bodies; larvae of the European *Osmylus fulvicephalus* Scopoli probe wet mud with their greatly elongated mouthparts, and feed on springtails (Collembola) in that habitat. Larvae of Kempyninae are more fully aquatic. Eggs are laid singly or in small groups, and may be inserted into grooves or holes in the substrate.

MAJOR REFERENCES — Krüger (1912-1915): comprehensive summary of early studies on family, and global synopsis of taxa; Tjeder (1957): South African taxa, with synopsis of most earlier work on family; Makarkin (1985): eastern Palaearctic fauna; New (1986b, 1988b, 1991a): treatments of Spilosmylinae from region.

Mantispidae (mantid flies)

54

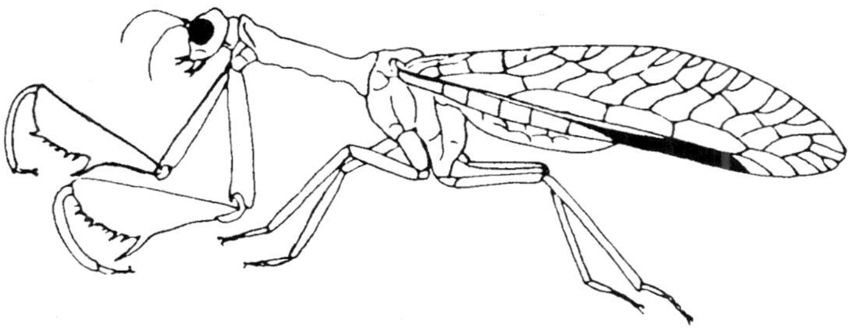

55

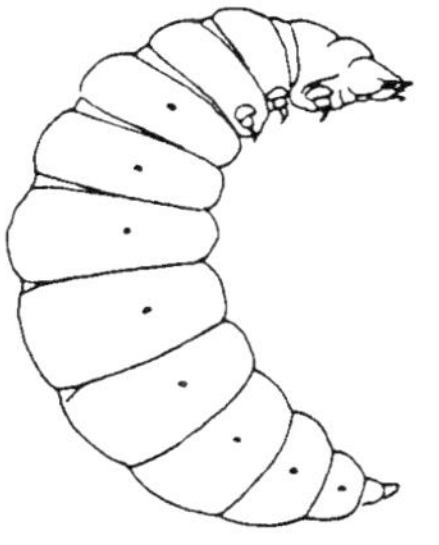

FIGURES 54 - 55
54 Adult Mantispidae
55 larval Mantispidae

Family diagnosis

Small to large lacewings (forewing length approximately 5-30 mm) with raptorial forelegs. Head rather small, vertex sometimes strongly domed; eyes large; ocelli absent; antennae short, flagellar segments simple to strongly broadened or discoidal; mandibles long; maxillary palpi five-segmented; labial palpi three-segmented. Prothorax elongate, tubular, sometimes strongly rugose and/or tuberculate; pterothorax usually broader than prothorax. Legs: coxa of foreleg greatly elongate, femur compressed laterally and with ventral spines, tibia arched and with lateroventral ridges; other legs long, slender, not elaborated; tarsi five-segmented; tarsal claw(s; one or two claws on fore tarsus) simple or bifid. Wings subequal, elongate, usually narrow and forewing slightly longer than hindwing; membrane hyaline to brightly coloured; trichosors usually absent; pterostigma well defined, usually dark (brown, black); veins generally bare, any macrotrichia usually inconspicuous. Costal crossveins few, simple; cell Sc sometimes almost obliterated due to contiguous nature of C and Sc; Sc and R1 linked by short crossvein behind pterostigma and one to three more basal crossveins; Sc and R1 reach wing margin separately; Rs and MA separate in basal third of wing, pectinate, with at least one series of gradate vein, linked to R1 by two or three crossveins; MA separates from Rs near base, pectinate; small jugal vein. Abdomen well sclerotised, usually shorter than wings.
Female: sternite vii sometimes modified with pouchlike excavation (crumena); subgenitale distinct; tergite ix narrow, entire or divided and with distinct ventral portion; gonapophyses laterales well developed; ectoprocts always separate dorsally, with small patch of trichobothria; spermatheca complex.
Male: sternite ix variable; ectoprocts usually distinct dorsally, cercal callus sometimes indistinct; gonarcus arched; mediuncus distinct, with slender

pseudopenis; parameres elongate; hypomeres sometimes present; hypandrium internum small.

Diversity and distribution

Mantispidae are very distinctive Neuroptera. There are at present about 420 species known in the world, of which 63 occur in Malesia. Their general appearance resembles that of small praying mantids, with which they are sometimes confused, in having strongly raptorial, spined forelegs and a greatly elongated prothorax. Amongst the more obvious features for differentiation, mantids (when winged) have a complex venation with many crossveins and a greatly enlarged hindwing, and conspicuous cerci. The characters of forelegs and prothorax are shared in Neuroptera only with Rhachiberothidae, which do not occur in Malesia. Adults frequently come to lights, but can also be captured by sweeping and beating vegetation, including tree blossoms.

Lambkin's (1986a) scheme for classification within the family has been adopted widely. Four subfamilies are separated as follows, but all the genera recorded from Malesia belong to one of these, Mantispinae. However, Drepanicinae and Calomantispinae occur in Australia and could possibly be found in the region.

Key to the subfamilies of Mantispidae (simplified from Lambkin, 1986a, b)

1 Pronotum in two pieces, without maculae ▸ **Symphrasinae** (New World only)
– Pronotum entire, strongly tubular, with maculae ▸ 2

2 Forelegs: coxa with a transverse sulcus at about 1/4-1/3 length; anterolateral margins of mesonotum acutely produced (mesonotal lobe; fig. 56) ▸ **Mantispinae**
– Forelegs: coxa without transverse sulcus; anterolateral margins of mesonotum rounded, not acutely produced (fig. 57) ▸ 3

3 Tarsal claws simple ▸ **Drepanicinae** (not yet found in Malesia)
– Tarsal claws apically bifid **Calomantispinae** (not yet found in Malesia)

FIGURES 56 - 61
Mantispidae - anterolateral margins of mesonotum **56** acutely produced or **57** rounded — **58** pronotal form in lateral view short and squat (*Euclimacia*), **59** profile 'stepped' (*Tuberonotha*), **60** profile regular (*Austroclimaciella*), or **61** pronotum long and slender (*Campanacella*) (58-61 from New, 1998).

56

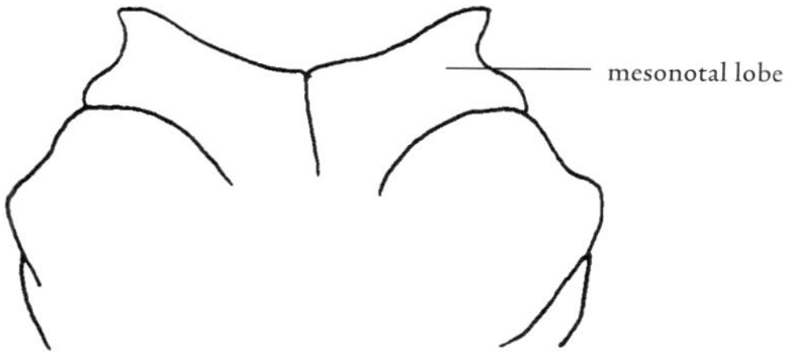

57

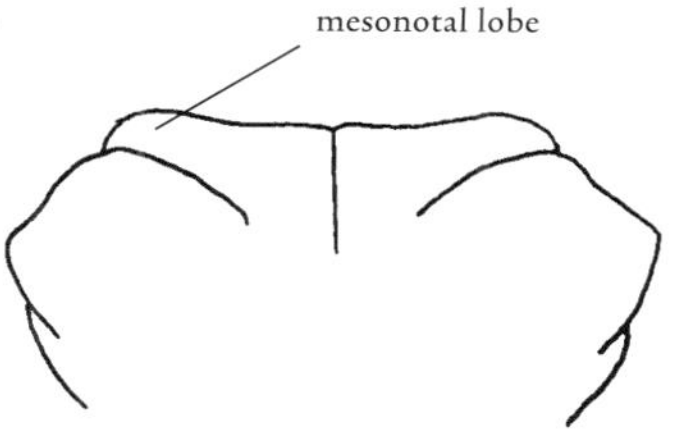

58

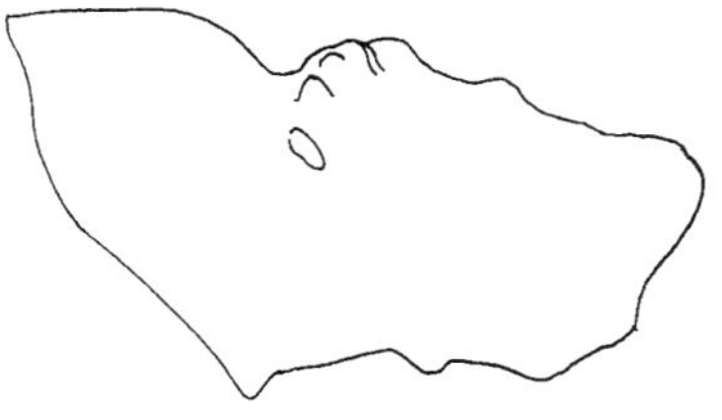

59

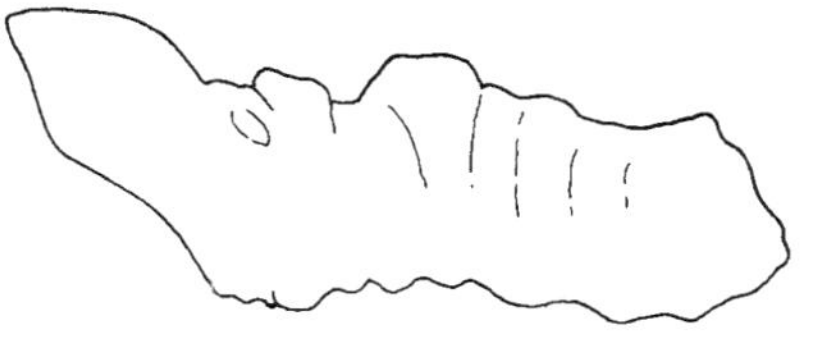

61

60

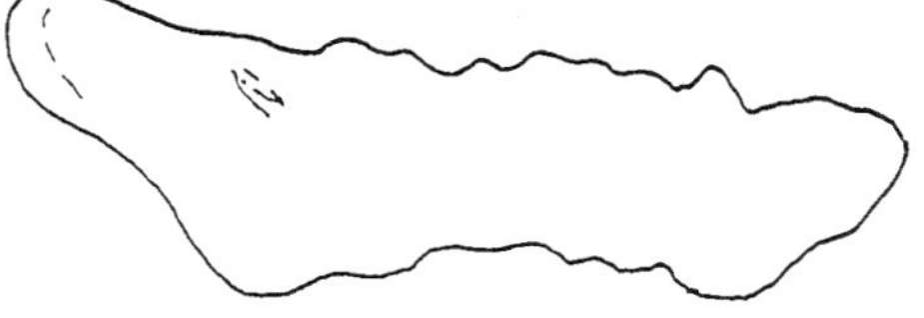

Additional diagnostic features of Mantispinae (after Lambkin, 1986b) include: tarsus of leg 1 with one claw and no arolium, and segment one at least as long as segments two-five together; tarsus of legs 2 and 3 with segment two longer than broad; no apical field of fine setae on medial face of tibia 1; forewing vein MP fused with R for considerable distance distal to crossvein 1M-Cu, and deflexed posteriorly at this crossvein to form a small triangular cell; forewing veins 2A and 3A fused basally to form thick common stalk; hindwing vein 2A short or absent, 3A absent; jugal lobe small.

The following key to genera is highly artificial, as only a limited range of characters have traditionally been used at this level. These are predominantly features of antennae (relative length, shape of flagellar segments), pronotum (shape, ornamentation) and wing venation. *Mantispa* Illiger s.l. is essentially a 'holding genus' to which many taxa have been referred uncritically; it undoubtedly includes species which will eventually form the basis of additional (currently undescribed) generic taxa. The values of genitalic features of both sexes were explored by Lambkin (1986a, b) for the Australian fauna, but those of most Malesian taxa have not yet been described formally. They are thus largely ignored in this account, other than for particularly conspicuous features visible without dissection (such as ectoproct shape). The major synthesis of the region's Mantispidae is by Handschin (1961), following the earlier generic level classification of Enderlein (1910).

Key to the genera of Mantispinae

(from New, 1998)

NOTE

The genera of Mantispinae known only from Australia (*Campion* Navás, *Toolida* Lambkin) are also included to aid interpretation of the New Guinea fauna, in particular. For pronotum shape, l:d ratio is the ratio of greatest length (l) to greatest depth (d) in lateral aspect.

1 Antennae perfoliate (segments near centre of flagellum strongly broadened, to around four times as broad as long) ▸ ***Euclimacia***
(prothorax very short [l:d ratio approximately 2.3:1], and 'humped' (fig. 58); forewing vein 2A forked; generally large species)

– Antennae normal (flagellar segments almost always at least as long as broad, usually distinctly longer) ▸ 2

2 Antennae long and slender, extending to more than half the length of the forewing ▸ ***Mimetispa***
(pronotum short [l:d ratio approximately 3.3:1]; forewing vein 2A forked; both wings with apex acute)

– Antennae relatively shorter, never extending beyond level of Rs separation in forewing ▸ 3

3 Prothorax relatively short and squat (l:d ratio $< 4:1$) ▸ 4

– Prothorax longer, generally slender (l:d ratio usually $> 4:1$) ▸ 7

FIGURES 62 - 63

Mantispidae - Mantispinae: representative wing venation: **62** condition of fused forewing veins C and Sc (*Austromantispa*), or **63** these veins clearly separate (*Tuberonotha*) (from New, 1998).

62

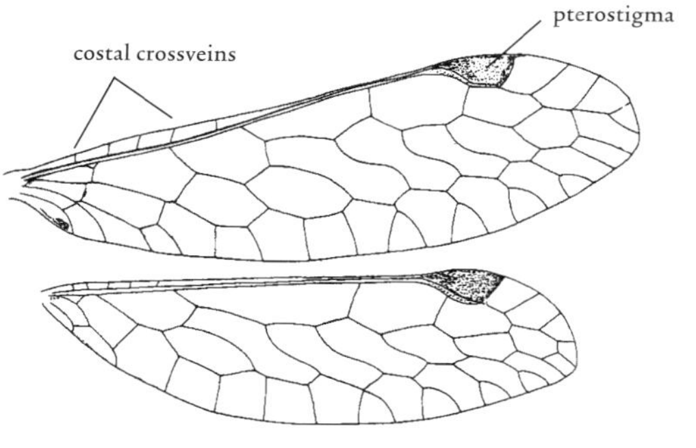

63

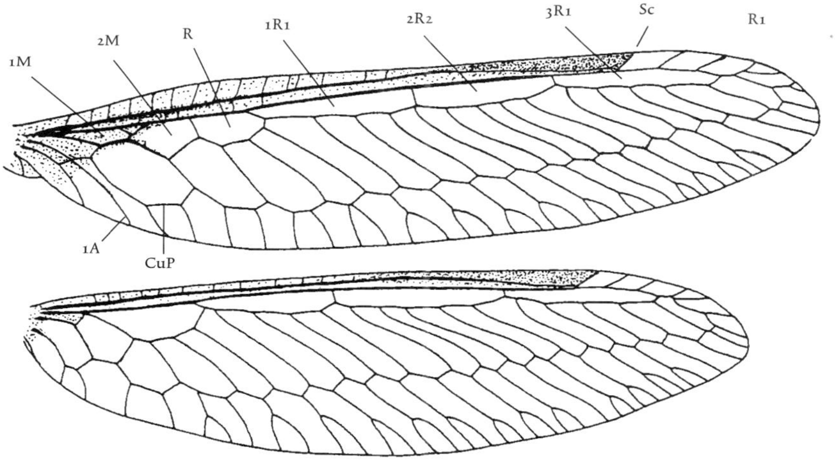

4	Abdominal tergites in central region raised dorsally to give pyramid-like appearance ▸	*Stenomantispa*
–	Abdominal tergites normal, not raised ▸	5
5	Pronotum with short thick setae on raised bases ▸ (male genitalia with accessory sclerites flanking gonocoxites)	*Campion*
–	Pronotum glabrous or with few very fine hairs, these never on raised bases	6
6	Pronotum distinctly 'stepped' in profile (fig. 59) ▸	*Tuberonotha*
–	Pronotum not as above, more regular in profile (fig. 60) ▸	*Austroclimaciella*

FIGURES 64 - 67
Mantispidae - diagnostic features of Mantispinae: male ectoprocts **64** elongate (*Orientispa*), or **65** not elongate (most others) — **66** forewing jugal lobe enlarged (*Toolida*), **67** forewing jugal lobe not enlarged (*Mantispa* s.l.).

64

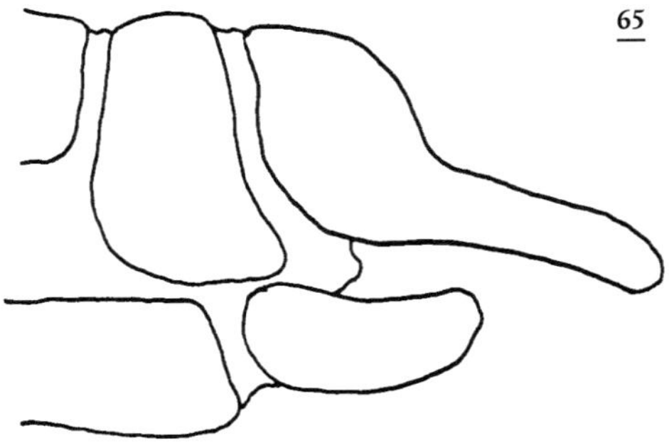

65

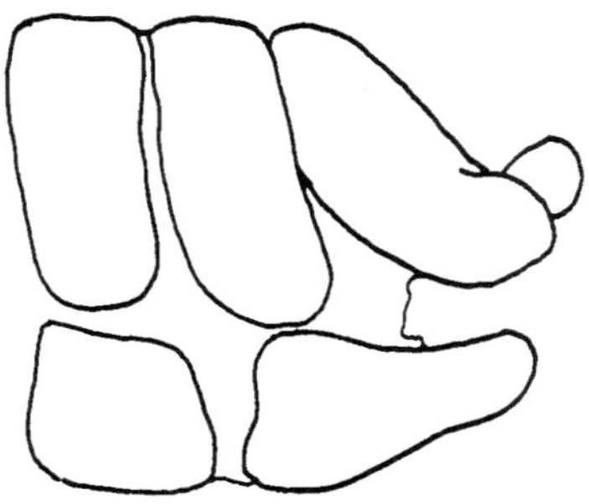

66

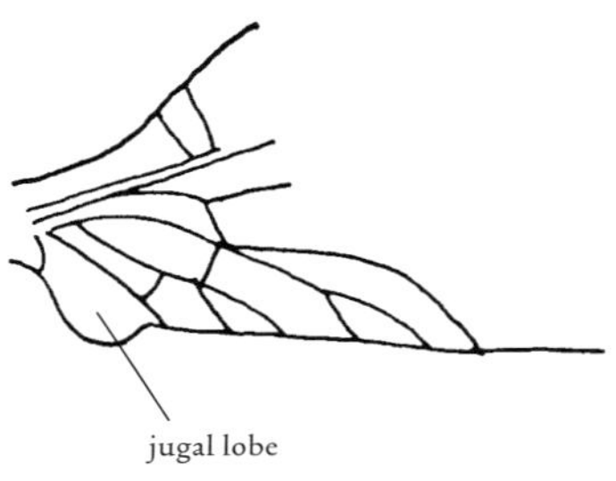

67

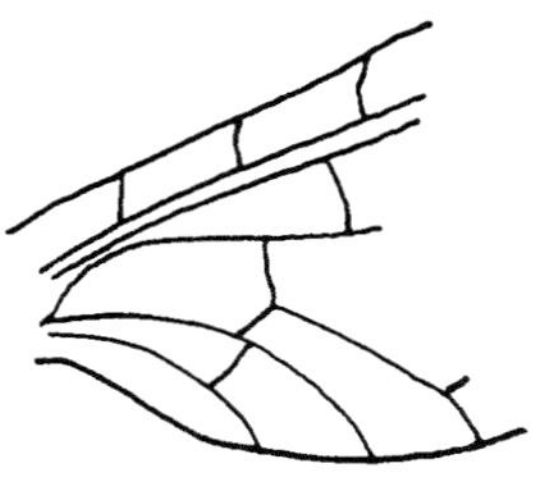

7	Forewing veins C and Sc fused basal to pterostigma (pterostigma usually short; fig. 62) ▸	8
–	Forewing veins C and Sc not fused (pterostigma usually long and shallow; fig. 63) ▸	9
8	Pronotum with short thickened setae ▸	*Austromantispa*
–	Pronotum glabrous ▸	*Xaviera*

(Some species of *Necyla* also key here, and many individuals are differentiated from *Xaviera* by having 1, rather than 2, veins leaving 2Ri. This character may be superficial and unreliable, and investigation is needed of this complex.)

9 One or more wings with supernumerary radial crossveins, so that more than three main R-cells are present ▸ *Eumantispa*
– No supernumerary radial crossveins ▸ 10

10 Forewing cell Cu1 asperous ventrally ▸ *Asperala*
– Forewing cell Cu1 normal, no thicker than rest of wing membrane ▸ 11

11 Large species (forewing length to 21–23 mm) ▸ *Campanacella*
– Smaller species (forewing length up to 15 mm, often 10 mm or less) ▸ 12

12 Male ectoprocts extended to form elongate 'claspers' (fig. 64) ▸ *Orientispa* (female not well defined)
– Male ectoprocts not conspicuously extended as above (fig. 65) ▸ 13

13 Pronotum with fine short hairs ▸ *Spaminta* (female sternite viii reduced and divided; male ectoprocts truncate)
– Pronotum glabrous or with short setae on raised bases ▸ 14

14 Forewing jugal lobe enlarged (fig. 66) ▸ *Toolida* (female sternite viii reduced but entire; male ectoprocts slightly extended)
– Forewing jugal lobe small or absent (as in fig. 67) ▸ *Mantispa, Mantispilla*

NOTE
some species allocated to *Necyla* (about 30 species allocated to these genera in the region; widespread; genitalia variable, and generic limits not defined for region)

Notes on the genera of Mantispinae

Asperala Lambkin (type species *Mantispa erythraea* Brauer). Two species: Indonesia, Papua New Guinea; also in Australia.

Austroclimaciella Handschin (type species *Mantispa quadrituberculata* Westwood) contains few described species, but the genus is reported from Indonesia, Papua New Guinea and the Philippines, in addition to Japan, China and (more tenuously) India.

Austromantispa Esben-Petersen (type species *Mantispa imbecilla* Gerstaecker) contains two species, from Indonesia and Papua New Guinea; also Australia.

Campanacella Handschin (type species *Mantispa hamiltonella* Westwood) contains only one species, from Indonesia and Malaysia.

Campion Navás (type species *Campion rubellus* Navás) includes eight species and is apparently restricted to Australia.

Euclimacia Enderlein (type species *Euclimacia partita* Enderlein) is one of the largest genera with around 25 species in the region (some of them undescribed), differentiated mainly on minor colour details.

Eumantispa Okamoto (type species *Mantispa suzukii* Okamoto) contains about eight regional species and is recorded from Indonesia, Papua New Guinea, Singapore. Occurs also in Taiwan, Japan, and possibly India.

Mantispa Illiger (type species *Mantispa pagana* Fabricius). As noted earlier, this is somewhat of a 'holding genus' and there is considerable confusion between *Mantispilla*, *Necyla* and *Mantispa* in the region.

Mantispilla Enderlein (type species *Mantispa indica* Westwood) is differentiated only poorly from *Mantispa* and is treated commonly as a synonym, as in the checklist.

Mimetispa Handschin (type species *Mantispa simulatrix* McLachlan) contains only one described species, from Indonesia and Malaysia.

Necyla Navás (type species *Necyla exigua* Navás) is also a member of the group of genera which are very similar to *Mantispa* and its status is unclear. In the checklist only those species placed formally in this genus rather than *Mantispa* are listed.

Orientispa Poivre (type species *Cercomantispa shirozui* Nakahara) contains one regional species, from Indonesia. The genus is known also from Japan and Sri Lanka.

Spaminta Lambkin (type species *Spaminta minjerribae* Lambkin) contains four species, from Indonesia, Papua New Guinea and Australia.

Stenomantispa Stitz (type species *Mantispa (Stenomantispa) ilsae* Stitz) contains two species, from Indonesia and Papua New Guinea. It was originally considered to be a subgenus of *Mantispa*, as implied above.

Toolida Lambkin (type species *Toolida infrequens* Lambkin). The genus contains only one described species, from Australia.

Tuberonotha Handschin (type species *Mantispa strenua* Gerstaecker). This striking genus contains around four species, and is widely distributed, being known from Indonesia, Papua New Guinea, Malaysia, and the Philippines, as well as from Japan and India.

Xaviera Lambkin (type species *Mantispa manca* Gerstaecker). The two species occur in Indonesia and Papua New Guinea; also Australia.

Biological outline

No details are available on the biology of Malesian Mantispidae. Some of the brightly coloured adults, at least some of them diurnal, appear to be mimics of vespoid wasps. It is almost certain that members of the genera noted above are all associated with spider hosts. Stange & Wang (1998) commented that all seven species from Taiwan (all in genera reported here) 'attack spider egg masses', but no more specific reference was provided. Unusually in Neuroptera, the larvae of this family undergo hypermetamorphosis, changing from an active triungulinoid first instar to vermiform later instars once a host is obtained. The high fecundity of some mantispids (Redborg & McLeod (1984) noted a female of a North American species laying 30,000 eggs!) may be associated with very high mortality levels on hatching: most larvae fail to find a spider egg sac or a female spider to board and await sac production.

MAJOR REFERENCES — Lambkin (1986a, b): revision of Australian fauna, with comprehensive global review of higher classification and relationships; Redborg & MacLeod (1984): biological studies and relationships with spiders.

Berothidae (beaded lacewings)

68

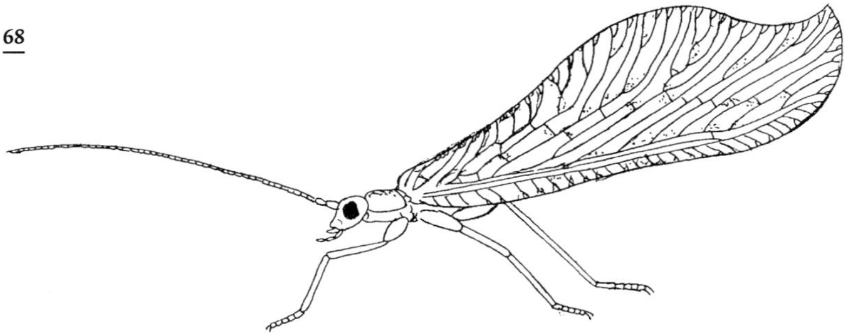

69

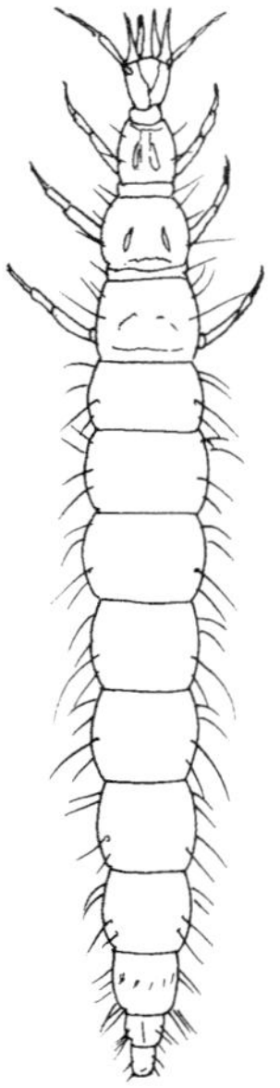

FIGURES 68 - 69
68 Adult Berothidae
69 larval Berothidae

Family diagnosis

Slender, small to medium sized lacewings (forewing 6-15 mm). Head rounded; ocelli absent; antennae moniliform, shorter than forewing, scape often enlarged; mandibles normal or reduced. Pronotum usually narrower than pterothorax, often slightly longer than broad. Legs cursorial, slender. Wings subequal, oval, narrow and straplike, or falcate; trichosors present; membrane hyaline or mottled; veins setose. Forewing: costal space narrow, many costal crossveins usually forked, humeral crossvein recurrent or simple; Sc and R1 distally fused at a point or linked by short crossvein; one basal Sc-R crossvein; two or three R1-Rs crossveins; CuA very long, parallel with hind margin of wing for much of wing length; CuP sometimes absent. Abdomen with tergites and sternites clearly defined; ectoproct and tergite ix distinct (female) or closely associated (male); cercal callus distinct.

Female: tergite viii usually deep; sternite vii large or reduced; lateral gonapophyses simple or with long hypocauda; subgenitale large or reduced; spermatheca usually complex and coiled.

Male: gonarcus complete; gonocoxites fused with it throughout their length or distally separated; mediuncus present, sometimes with pseudopenis at apex.

Sexual dimorphism can occur in eye size, shape of head, length of antennae, and wing ornamentation.

Diversity and distribution

About 80 species of Berothidae have been described in the world and six occur in Malesia. The Berothidae appear to be closely related to Mantispidae, but are themselves complex, with a number of distinct subfamilies. In many classification schemes, four main subfamilies are recognised, with the former family Rhachiberothinae now accorded family rank as Rhachiberothidae (Aspöck & Mansell, 1994), a sister group to the true Berothidae. MacLeod & Adams (1967) established the higher classification of Berothidae used by many later workers, pending separation of Rhachiberothidae. The conventional arrangements were changed somewhat in a cladistic analysis of the family by Aspöck & Nemeschkal (1998), in which the authors recognised four monophyletic groups, formalised as subfamilies Protiobiellinae, Trichomatinae, Nosybinae and Berothinae, and an unresolved trichotomy of the genera *Nyrma*, *Cyrenoberotha* and *Manselliberotha* which collectively constitute the former subfamilies Nyrminae and Cyrenoberothinae.

The most diverse subfamily, Berothinae, still contains eleven genera after transfer of another six to comprise Trichomatinae, Protiobiellinae and augment Nosybinae (two genera each), and is the only group of Berothidae known from Malesia. In the earlier arrangement of MacLeod & Adams (1967), Berothinae were separated from Cyrenoberothinae by abdominal tergite ix fused with the ectoproct (rather than free) and the face very short and not extending ventrally far below the lower margin of the eyes (rather than extending far more conspicuously ventrally). They were separated from Nosybinae by the separate bases, or at least distinct origins of forewing veins R and M (rather than these veins being fused, with a common stem) and the short female sternite viii (rather than this being an elongate triangular structure extending ventrally below more posterior structures).

Notes on the genera

Many of the genera are now reasonably distinct (Aspöck, 1986), but some changes were made or confirmed by Aspöck (1986), and the status of some genera in the region formerly treated as valid (e.g. *Acroberotha*, now a synonym of *Isoscelipteron*) or synonyms (e.g. *Isoscelipteron* with *Berotha*, [such as by Tjeder, 1959] now treated as distinct) has become clearer. The species from Malesia are generally sparse in collections but it is notable that many of the genera found in northern Australia appear to be absent. A simple key to the putative genera in the region is at present difficult to construct. *Berotha* and *Isoscelipteron* were noted as being among the 'higher' Berothinae by Aspöck & Nemeschkal (1998), who did not recognise *Frawalkeria* in their character analysis. Generic separation in this complex is mainly on small details of genitalia, and specimens should be submitted to a specialist for accurate determination.

Berotha Walker (type species *Berotha insolita* Walker) is a small genus (Aspöck, 1983) with one species (*B. piepersii* van der Weele) described from Java and known also from Singapore, and others from India and Sri Lanka. It was formerly somewhat of a 'holding genus' to which a variety of species were referred uncritically. Following Aspöck's (1983) careful reappraisal, and definition on genitalic features, *Berotha* is now known to be restricted to the above region and does not occur in Europe or Africa, as was previously supposed.

Isoscelipteron Costa (type species *Isoscelipteron fulvum* Costa; figs. 70-72) has the widest distribution of any berothid genus, but comprises rather few species (Aspöck & Aspöck, 1983, 1991). About four have been described from the region. *I. nicobaricum* (Navás) is widely distributed; *I. leveri* (Kimmins) occurs in the Solomon Islands; *I. borneole* Aspöck & Aspöck in Borneo, and *I. philippinicola* Aspöck & Aspöck was described from the Philippines. All appear to be rare, and few specimens are known. *Dasypteryx* Stein and *Acroberotha* Krüger are synonyms.

Frawalkeria Navás (type species *Isoscelipteron puncticollis* Navás) needs critical reevaluation to determine its validity. It was considered a synonym of *Berotha* by Tjeder (1959), but has been treated more recently as incertae sedis.

Biological outline

Very few details of berothid biology are available from any part of the world. Adults readily come to light, but few developmental stages are known. One North American species is an obligate termitophile, and the larvae produce an allomone that immobilises termites on which they feed. It has been assumed widely that such specialised habits may be more widespread in the family, but there is no direct evidence for this. As for many other Neuroptera, information on the Malesian taxa would be of considerable interest in augmenting the meagre information available from elsewhere.

MAJOR REFERENCES — Tjeder (1959): South African fauna, and summary of much earlier work; MacLeod & Adams (1967): higher classification and definition of traditional subfamilies of world fauna; Johnson & Hagen (1981): biological outline; first major life history information on family; Aspöck (1986): synopsis of family.

FIGURES 70 - 72
Berothidae - Berothinae: **70** fore- and hindwings of *Isoscelipteron* — **71** schematic abdominal apex of *Isoscelipteron*, lateral aspect of male, and **72** of female.

70

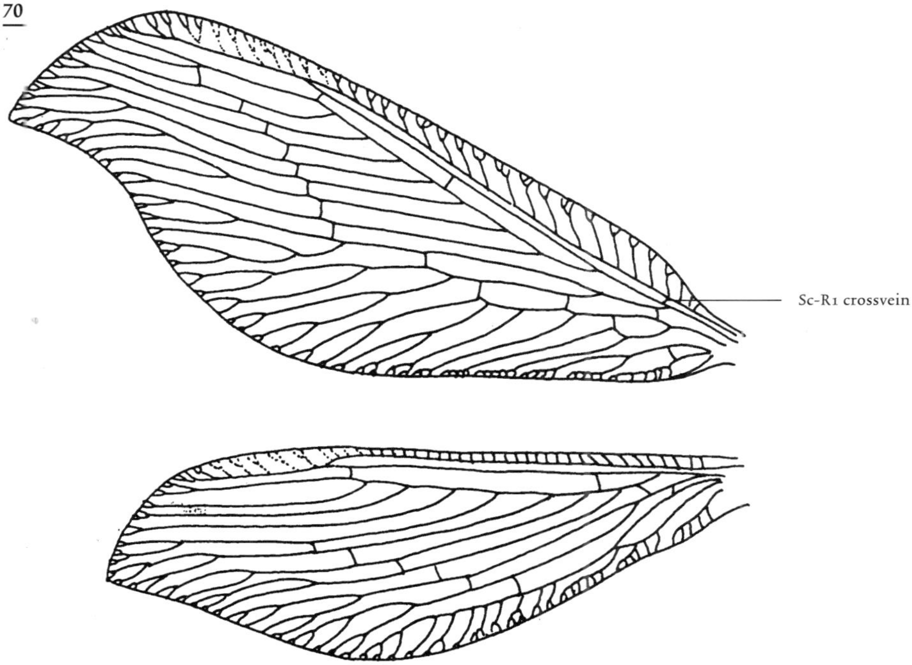

72

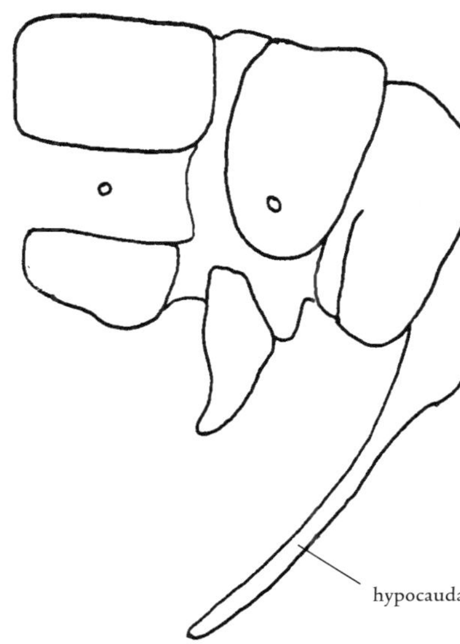

71

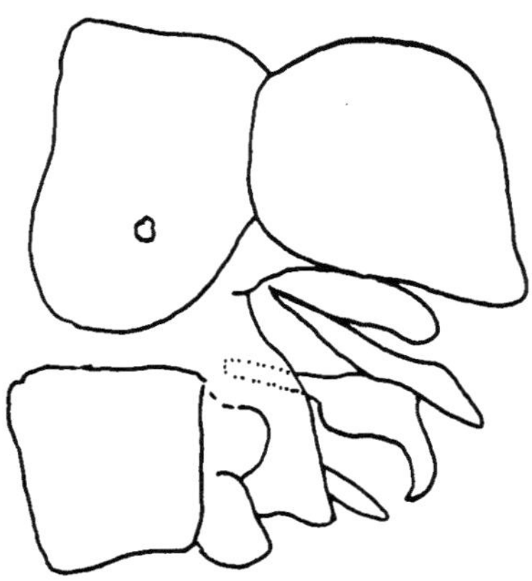

Dilaridae (pleasing lacewings)

73

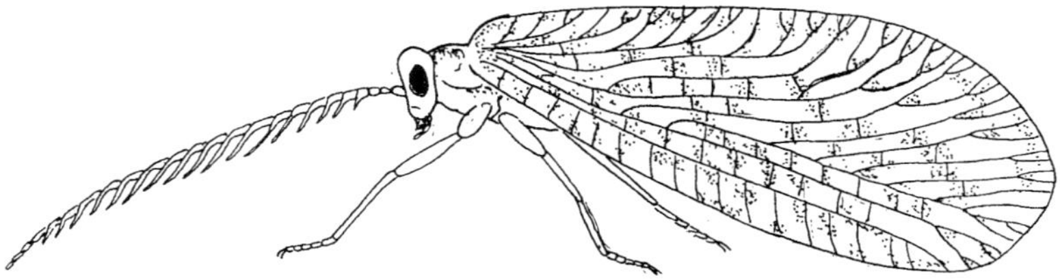

74

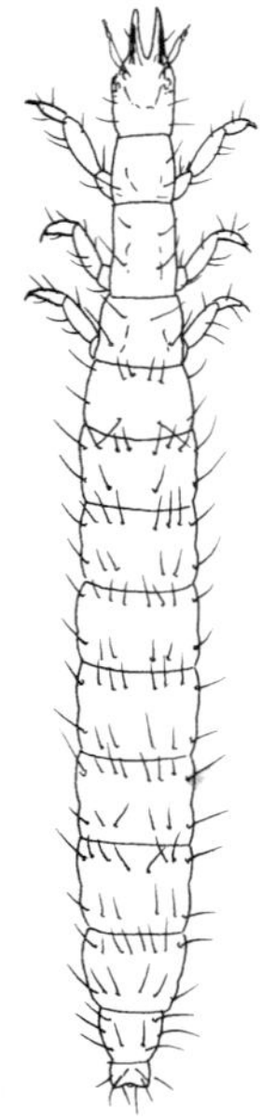

FIGURES 73 - 74
73 Adult Dilaridae
74 larval Dilaridae

Family diagnosis

Small to medium sized lacewings (forewing length approximately 4-12 mm), delicate. Eyes large; ocelli represented by raised tubercles; antennae pectinate in male (fig. 73, 75), filiform in female; mouthparts strongly reduced. Prothorax at least as wide as long, usually wider; pterothorax broad. Legs slender, unornamented. Wings (fig. 76) subequal, densely hairy, membrane sometimes mottled; sexually dimorphic: forewing of male relatively broader and shorter than that of female, hindwing usually narrower than forewing in female, sometimes broader in male. Forewing: costal space narrow, costal crossveins simple or a few forked, no recurrent humeral vein; pterostigma not defined; trichosors present; R1 and Rs separate to wing margin; one to three R1-Rs crossveins; Rs with two to five branches, with few (if any) crossveins, sometimes in a single gradate series; MA arises separately (origin variable: figs. 77-79); MP deeply forked or 'twigged' near wing margin. Abdomen short, moderately stout.

Female: ectoproct small, with cercal callus; tergite ix deep; sternite viii often reduced; lateral gonapophyses form long ovipositor, extending for several mm; posterior gonapophyses short; two globular spermathecae.

Male: ectoproct transverse, often with hooked lobes (Nallachiinae) or reduced with compensatory enlargement of tergite ix (Dilarinae); cercal callus defined; tergite ix deep; gonarcus arched; parameres long and slender; hypandrium internum small.

FIGURES 75 - 79

Dilaridae - **75** male antenna with lateral processes on flagellar segments — **76** fore- and hindwings of *Dilar* — **77** different conditions of association of forewing veins R, Rs and MA in *Rexavius marmoratus*, MA and R connected only by crossvein (XV, dotted), **78** MA appearing as a basal branch (b) from R (other Dilarinae), **79** MA appearing as branch from Rs (Nallachiinae, not in Malesia) (77-79 after Adams, 1970).

75

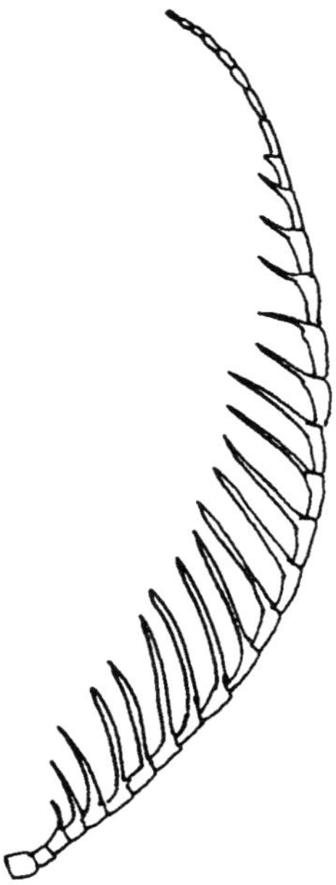

76

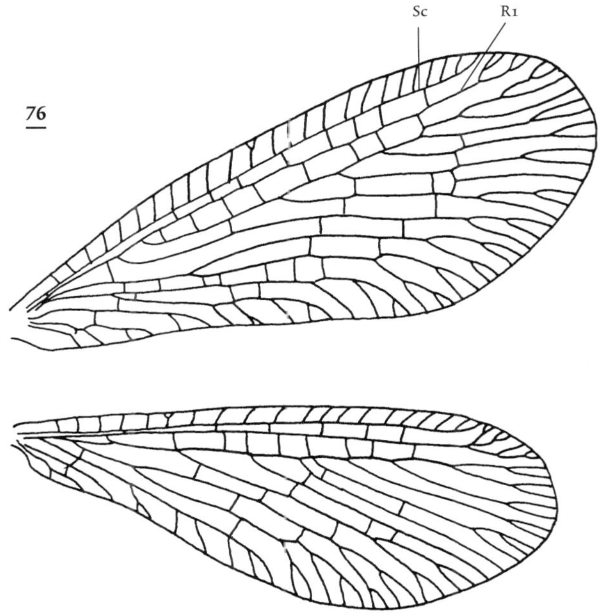

77

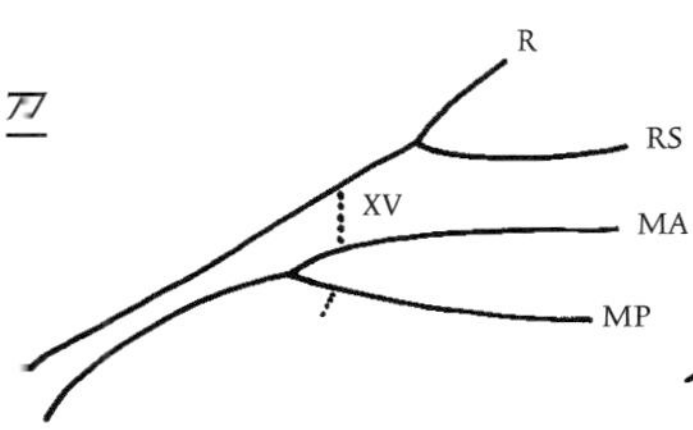

78

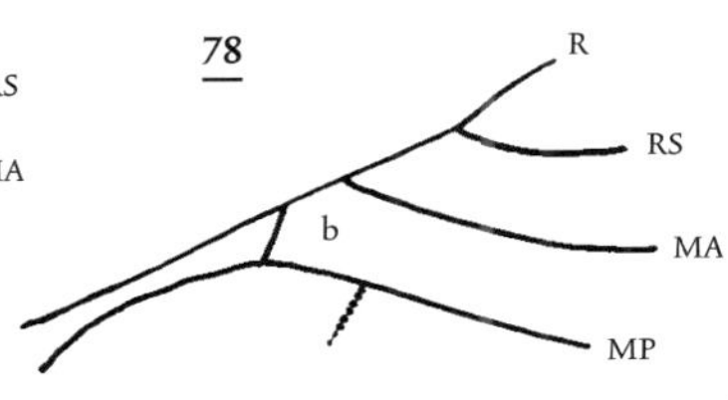

79

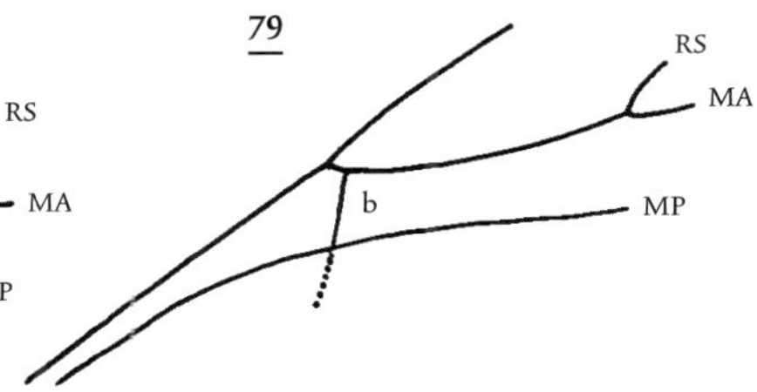

Diversity and distribution

Dilaridae is a small family, with only about 40 described species in the world, divided between two subfamilies. Three species have been found in Malesia. The family is absent from Australia, and the small Malesian representation apparently reflects attenuation from the more westerly Palaearctic faunas. The subfamily Nallachiinae is largely confined to the New World, with one anomalous species in southern Africa. The subfamily Dilarinae is widespread.

Key to the subfamilies of Dilaridae

1 Forewing vein MA separating from stem of R before separation of R1 and Rs, rarely more distinctly separate at wing base (figs. 77, 78); male antennae with several apical segments lacking a lateral process (fig. 76) ▸ **Dilarinae**

– Forewing vein MA arising as branch from Rs (fig. 79); male antennae never with more than three apical segments lacking lateral processes ▸ **Nallachiinae** (not in Malesia)

Larvae of very few species of the family are known. Those of Dilarinae appear to be associated with soil, and those of Nallachiinae with dead wood (see Macleod & Speigler, 1961; Minter, 1996).

Notes on the genera of Dilaridae

Berothella (type species *Berothella phantoma* Banks) is a small genus with two described species from Malaysia and China, respectively.

Dilar Rambur (type species *Dilar nevadensis* Rambur) is widely distributed in the Palaearctic, from southern Europe to Japan, Taiwan and China, and in northern Africa. It is represented in Malesia by two species transferred from *Rexavius* and by a recently described species from Borneo. The former genus *Rexavius* Navás (type species *Dilar nietneri* Hagen) contained three species, one (with two nominal subspecies) from Japan, and the others from Sri Lanka and Thailand. It was considered a synonym of *Dilar* by Kuwayama (1962), but has been treated as distinct by some later workers. However, two species (*grandis* Banks, *marmoratus* Banks) from the region were transferred to *Dilar* by Oswald (1998b), who confirmed the synonymy of the genus with *Dilar*.

Biological outline

Biological details are not available for any Malesian dilarid.

MAJOR REFERENCES — Adams (1970): review of systematics of family; Oswald (1998b): catalogue of world species.

Hemerobiidae (brown lacewings)

80

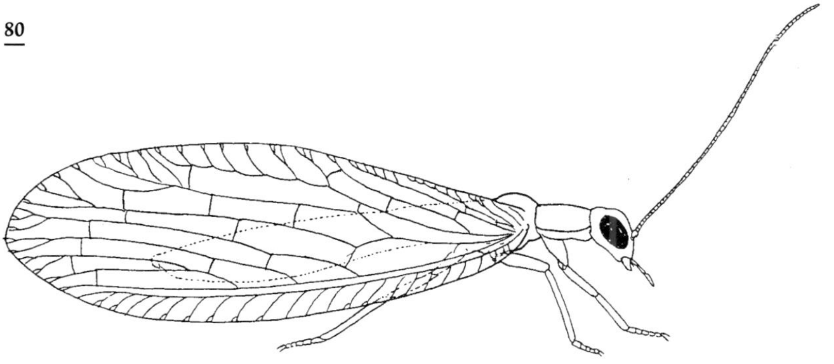

81

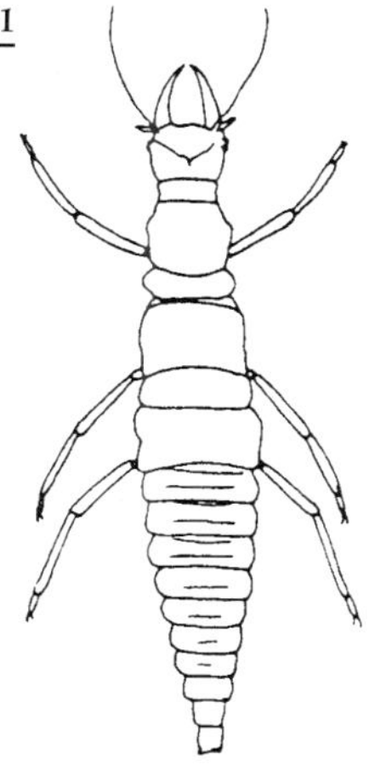

FIGURES 80 - 81
80 Adult Hemerobiidae
81 larval Hemerobiidae

Family diagnosis

Small to medium sized delicate lacewings (forewing approximately 3-18 mm, mostly 4-10 mm). Head small to medium; ocelli absent; antennae slender, generally moniliform to filiform, antennal length approximately 1/2 to >1 times forewing length, scape enlarged; mandibles strong; maxillary palpi five- or six-segmented; labial palpi three- or four-segmented. Pronotum usually short, commonly broader than long; mesonotum larger than metanotum. Legs slender, cursorial; tibiae sometimes expanded or with short spurs; tarsal claws simple. Wings subequal, usually oval, more rarely elongate or falcate; hindwing sometimes reduced; membrane hyaline to strongly patterned; trichosors present; pterostigma usually distinct. Venation variable but with unifying feature of forewing Rs having more than one (usually two to four) branches from R1; forewing costal space variable, humeral crossvein simple or recurrent; gradate veins variable but usually few series; forewing with jugal lobe. Abdomen slender; tergites and sternites well developed; cercal callus and trichobothria present.

Female (fig. 82): tergite ix broadened ventrally below ectoproct; tergite viii extended ventrally; lateral gonapophyses prominent, rounded, sometimes with apical style; subgenitale usually present; spermatheca simple.

Male (figs. 83, 84): posterior tergites usually simple, but occasionally with dorsal posterior lobes; tergite ix sometimes fused with ectoproct, rarely with long lateral process; sternite ix short; ectoproct often elongated, sometimes more

FIGURES 82 - 84

Hemerobiidae: generalised terminology of abdominal apex in lateral aspect of **82** female, and **83** male — **84** details of arrangement of male genitalia in lateral aspect (upper), hypandrium internum and parameres in ventral aspect (lower left) and gonarcus and arcessus in dorsal aspect (right) (from New, 1988a).

82

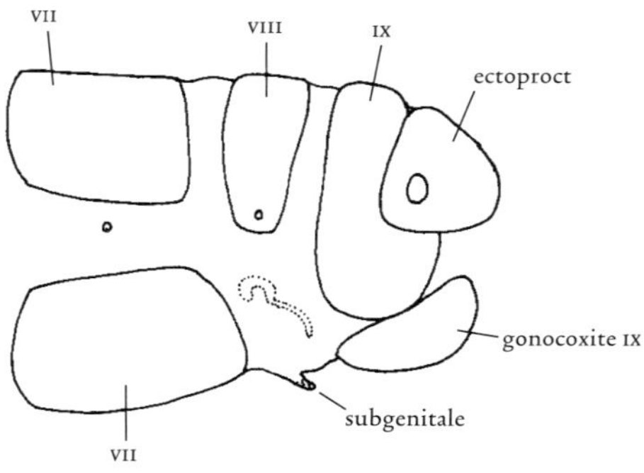

83

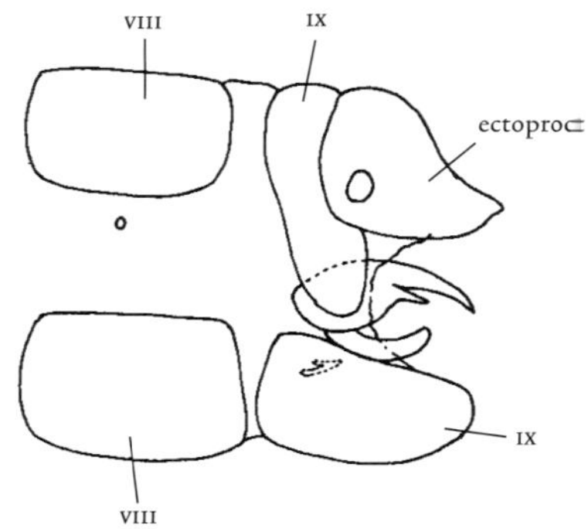

84

(**g**) gonarcus
(**a**) arcessus
(**p**) parameres
(**hi**) hypandrium internum

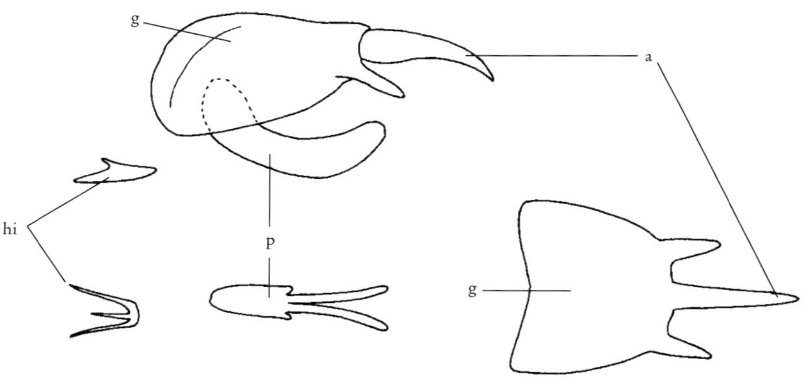

extensively ornamented; gonarcus transverse; entoprocessus sometimes present; parameres simple, free or fused; superprocessus present or absent; hypandrium internum small.

Classification of Hemerobiidae is mainly based on features of wing venation and genitalia, the complex structures of the male genitalia often being particularly characteristic. Subfamilies are more difficult to define simply, because of the mosaic nature of the features on which some are characterised, but much of the character suite relies on wing venation. Many of the larger genera, likewise, have series of synonyms, some of which represent attempts to divide large, unwieldy taxa by designation of subgenera, often geographically restricted and raised on rather small features of individual species. Oswald (1993a) suggested that formal division at this level was premature, as many of the supposed taxa were not demonstrably holophyletic. Thus I have simply listed names as putative generic synonyms, because they occur in regional literature on the family, and have not accorded the genera any other formal status, although indicating the subfamily placement suggested by Oswald.

Diversity and distribution

The species of Hemerobiidae are distributed amongst 26 described genera in 10 subfamilies (Oswald, 1993a, 1994). Many of the higher taxa are very limited in distribution, but some genera (such as *Hemerobius* Linnaeus, *Notiobiella* Banks and *Micromus* Rambur s.l.) are widespread. The last-named in particular occurs very widely in Malesia. A number of more restricted genera represented in Indonesia and Papua New Guinea are related to the Australian fauna (New, 1988a, 1989b). Hemerobiidae are a cosmopolitan family with around 550 described species, of which 50 have been recorded from Malesia.

Key to the genera of Hemerobiidae (after Oswald, 1993a)

NOTE
The following key to genera includes a few taxa not yet reported from Malesia but which are known from adjacent regions and might be found there in due course.

1 Forewings with two or fewer apparent Rs veins ▸ 2
– Forewings with three or more apparent Rs veins ▸ 6

2 Forewing vein CuP forked basal to crossvein 2CuA-CuP (fig. 85) ▸ ***Notiobiella***
– Forewing vein CuP not forked basal to crossvein 2CuA-CuP ▸ 3

3 Posterior sectoral trace simple along proximal half (fig. 86) ▸ ***Carobius***
– Posterior sectoral trace with one side branch along posterior half (fig. 87) ▸ 4

4 Forewing costal space very narrow; posthumeral costal crossveins simple; no trichosors in humeral area (fig. 88) ▸ ***Zachobiella***
– Forewing costal space narrow or broad; usually some posthumeral costal crossveins forked or branched (rarely, none branched); trichosors in humeral area ▸ 5

5 Outer gradate series of forewings (cf. fig. 89) with three or more (commonly four) intraradial crossveins ▸ ***Sympherobius***
– Outer gradate series of forewings with no more than two (usually, none) intraradial crossveins (fig. 87) ▸ ***Psectra***

FIGURES 85 - 91
Hemerobiidae: wing venation, fore- and hindwings of representative **85** *Notiobiella*, **86** *Carobius*, **87** *Psectra*, **88** *Zachobiella*, **89** *Micromus*, **90** *Drepanacra* — **91** forewing venation, terminology (*Micromus*).

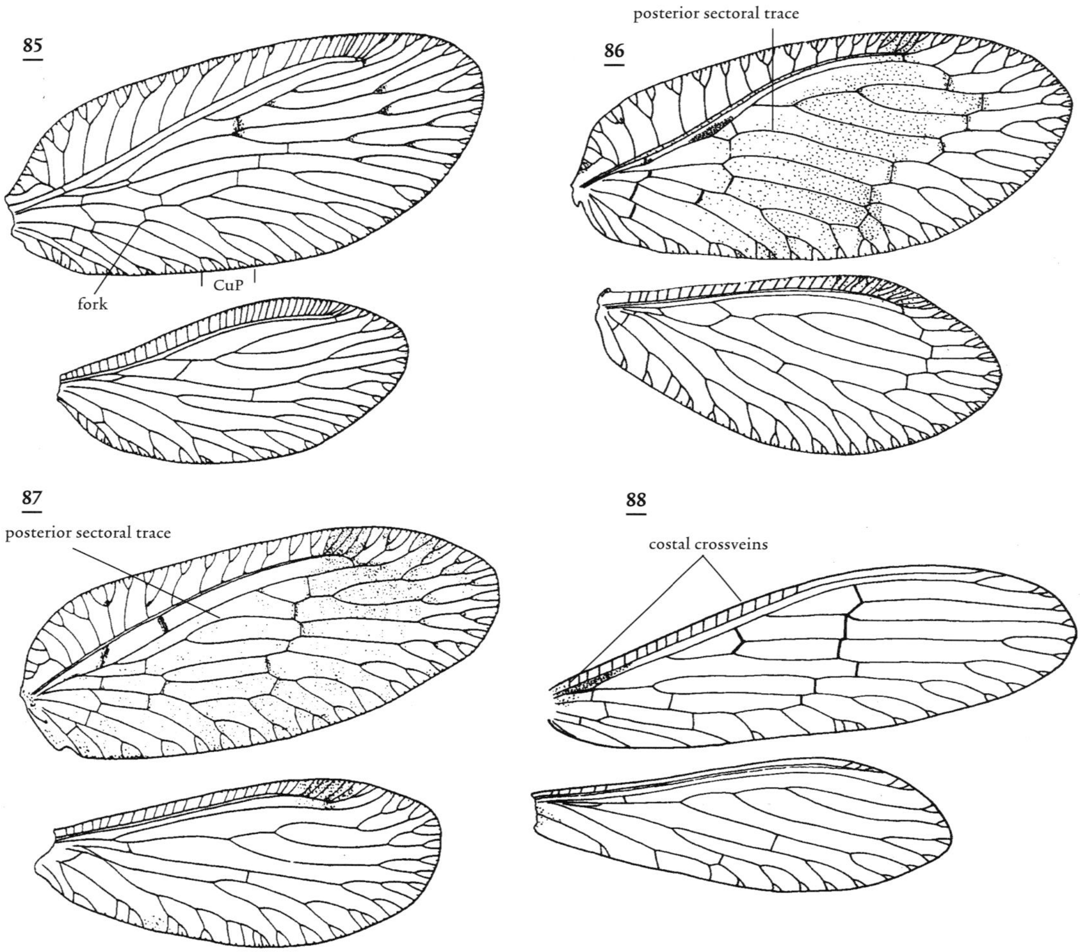

6 Both forewings with two or more prestigmal Sc-R crossveins (fig. 90) ▸ 7
– Both forewings with one (rarely, none) prestigmal Sc-R crossveins ▸ 9

7 Forewings falcate (fig. 90) ▸ *Drepanacra*
– Forewings not falcate ▸ 8

8 Forewings with three or more intraradial crossveins ▸ *Neuronema*
– Forewings with two or less intraradial crossveins ▸ *Megalomus*

89

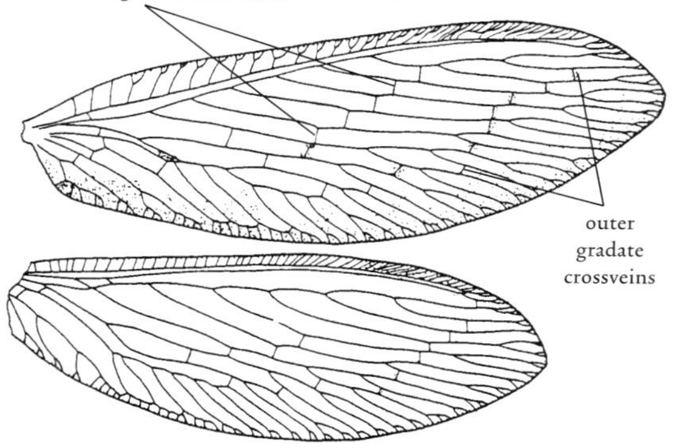

90

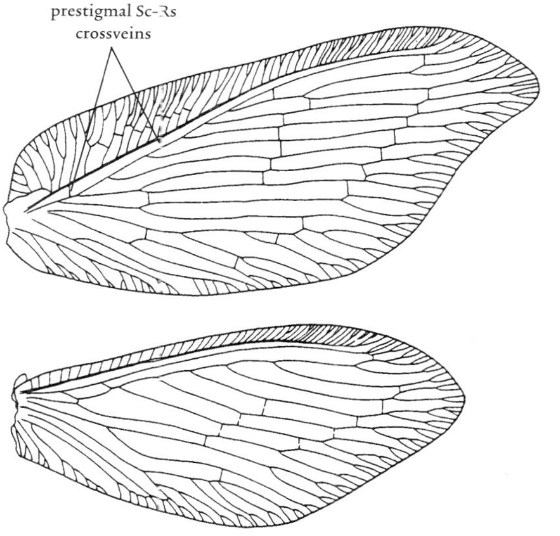

91

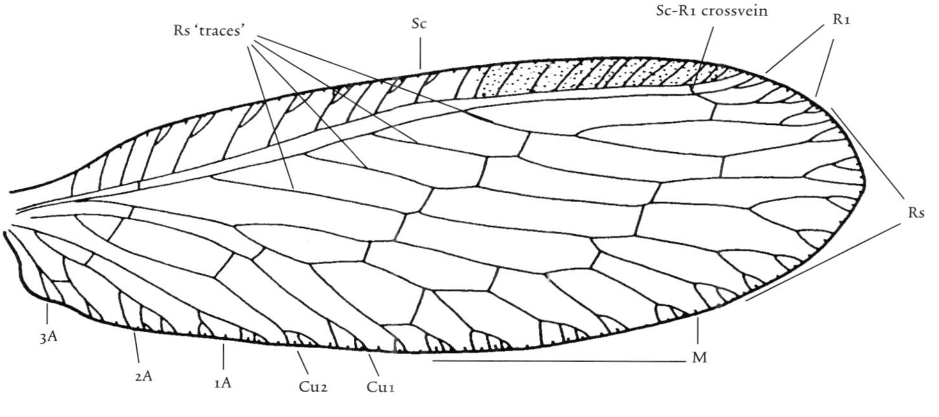

9	Recurrent humeral vein (fig. 9) absent ▸	10
–	Recurrent humeral vein (fig. 8) present ▸	11
10	Forewing crossvein 2M-Cu present ▸	*Noius*
–	Forewing crossvein 2M-Cu absent (figs. 89, 91) ▸	*Micromus*
11	Forewings strongly tapered to acute apex ▸	*Megalomina*
–	Forewing apex broader, rounded or with only slight indication of acute apex ▸	12

FIGURES 92 - 95
Hemerobiidae: clypeus of **92** *Hemerobius* and **93** *Wesmaelius*, to show characteristic sagittal seta of *Hemerobius* (after Oswald, 1993a) — **94** fore- and hindwings of *Hemerobius*, and of **95** *Wesmaelius*.

92

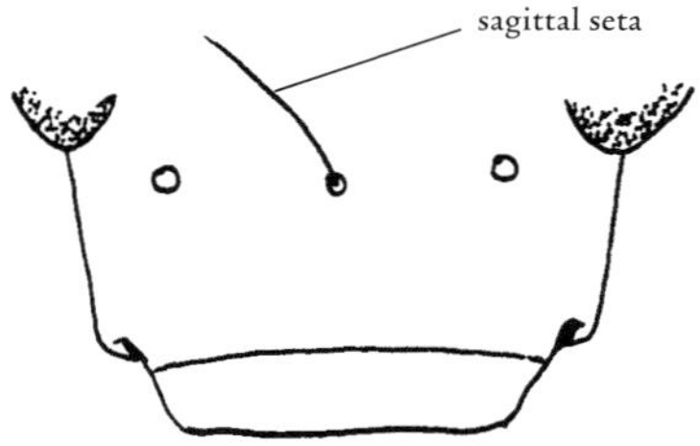

93

94

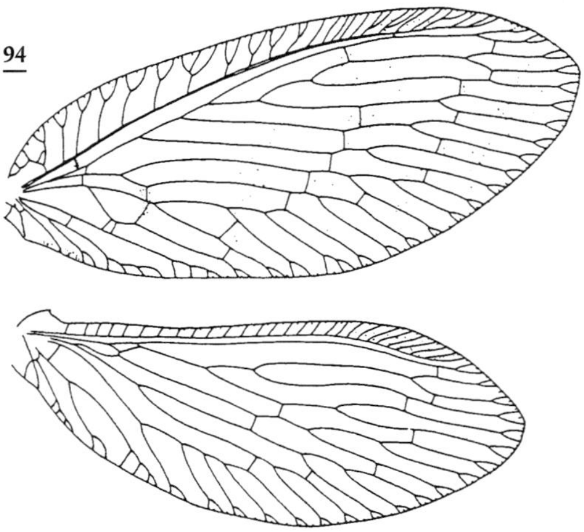

95

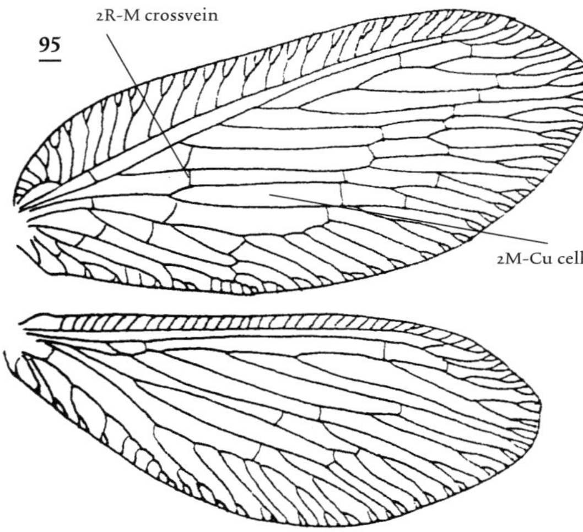

12 Hindwings with outer gradate series with one or fewer (rarely, two) intraradial crossveins ▸ ***Sympherobius***

– Hindwings with outer gradate series with three or more (rarely, two) intraradial crossveins ▸ 13

13 Clypeus with long sagittal seta (fig. 92); crossvein 2R-M usually absent (if present, proximal to crossvein 2M-Cu); cell 2M-Cu narrow distally (fig. 94) ▸ ***Hemerobius***

– Clypeus without long sagittal seta (fig. 93); crossvein 2R-M present, distal to crossvein 2M-Cu; cell 2M-Cu broad distally (fig. 95) ▸ ***Wesmaelius***

Notes on the genera

Hemerobius Linnaeus (type species *Hemerobius humulinus* Linnaeus; Hemerobiinae; figs. 94, 96, 97) is a large, cosmopolitan genus with about 125 described species. Somewhat inexplicably, it is poorly represented in much of Southeast Asia, but one species occurs in Australia, and three have been reported from New Guinea (New, 1989b). It appears to be more diverse in the Philippines, but the nominal species from there are in need of revision. *Hemerobius* is most likely to be confused with *Wesmaelius* (below), but differs in some small venational details and in the clypeus having a long sagittal seta at about the level of the anterior tentorial pits (Oswald, 1993a). Male genitalia of the two genera also differ: the parabaculum of *Hemerobius* is completely divided, and the ectoproct lacks a pecten.
There exist many generic synonyms, reflecting the wide distribution of *Hemerobius*; they include *Mucropalpus* Rambur, *Hemerodromia* Navás, *Brauerobius* Krüger, *Anotiobiella* Kimmins, *Dyshemerobius* Tjeder, *Mesohemerobius* Nakahara and *Semohemerobius* Yang.

Notiobiella Banks (type species *Notiobiella unita* Banks; figs. 85, 98, 99) is closely related to the following two genera, all placed in Notiobiellinae. It is widely distributed in the region. A number of the species have very broad, bluntly rounded wings which are held virtually flat whilst at rest, rather than in the more usual vertical 'tent' posture. *N. viridis* Tillyard is one of very few bright green hemerobiids. *Vaja* Navás, *Buxtonia* Esben-Petersen and *Ganchetus* Navás are synonyms.

Psectra Hagen (type species *Hemerobius diptera* Burmeister; figs. 87, 100, 101) differs from *Notiobiella* mainly by having forewing vein CuP forked basal to crossvein 2CuA-CuP. It closely resembles the African *Anapsectra* Tjeder (known from two described species), and differs from it mainly on male genitalic features. *Psectra* appears to be widely distributed in Malesia, but the species are generally not common. Synonyms include *Annandalia* Needham, *Eucarobius* Esben-Petersen and *Kimminsiella* Nakahara.

Zachobiella Banks (type species *Zachobiella punctata* Banks; figs. 88, 102, 103) shares the above CuP fork position with *Psectra*, but differs substantially from other notiobielline genera by having the forewing very narrow and with short, unbranched costal crossveins. Male genitalia are elaborate and the posterior tergites sometimes have elongated processes. The genus is restricted to Southeast Asia and Australia, and at least one species (*Z. pallida* Banks) is quite

FIGURES 96 - 105

Hemerobiidae - appearance of abdominal apex (male, female in sequence, lateral aspect): ***96, 97*** *Hemerobius*, ***98, 99*** *Notiobiella*, **100, 101** *Psectra*, **102, 103** *Zachobiella*, **104, 105** *Micromus*.

96

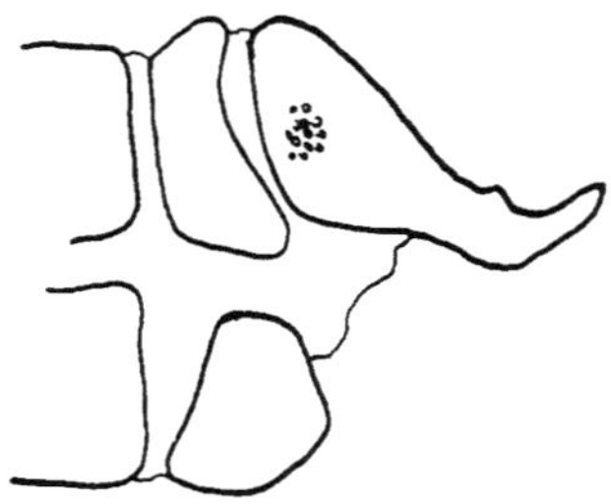

97

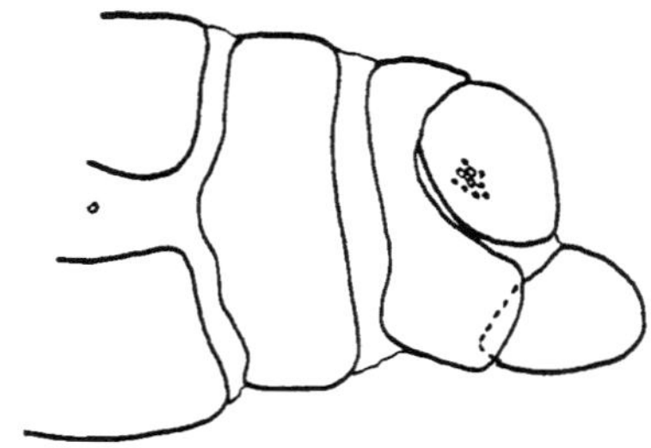

98

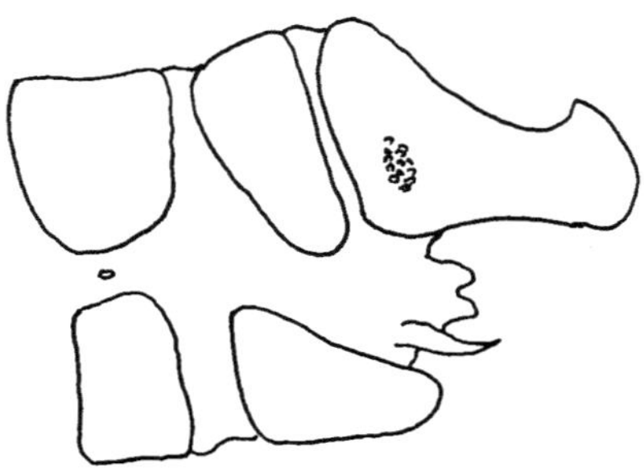

99

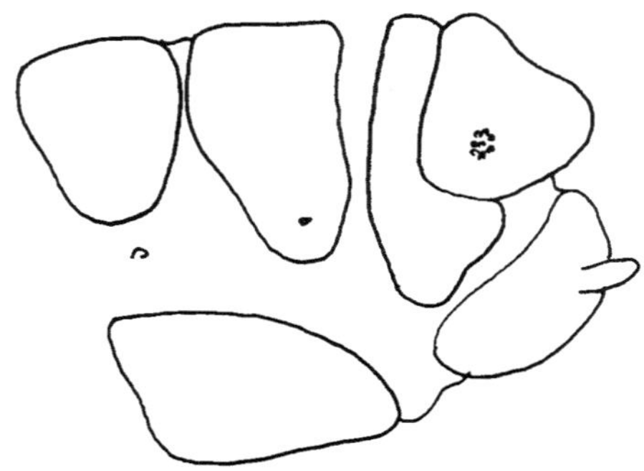

100

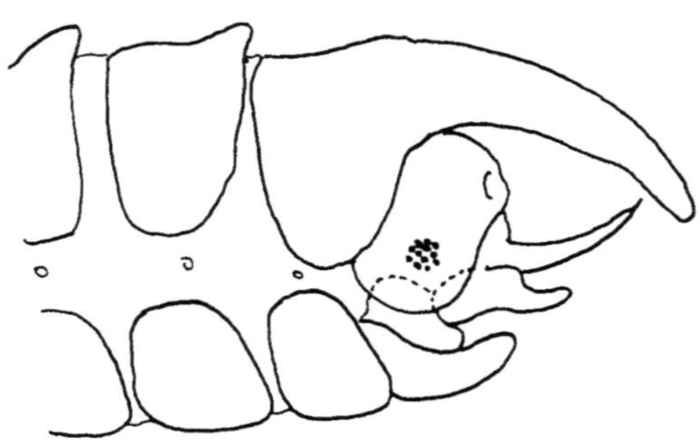

101

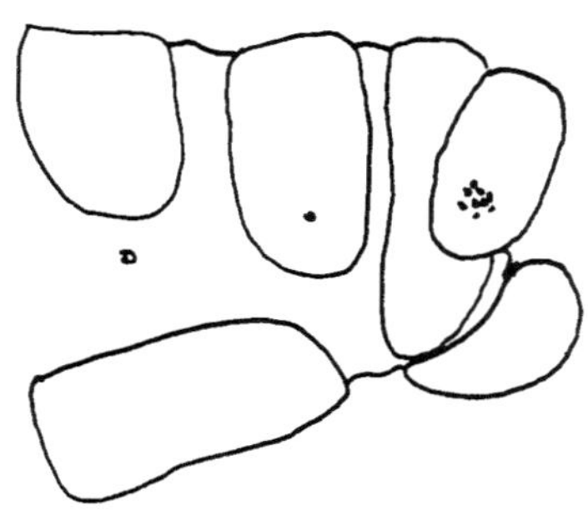

102

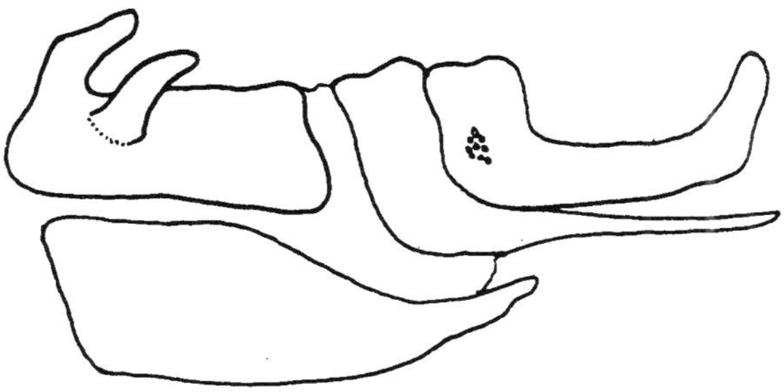

103

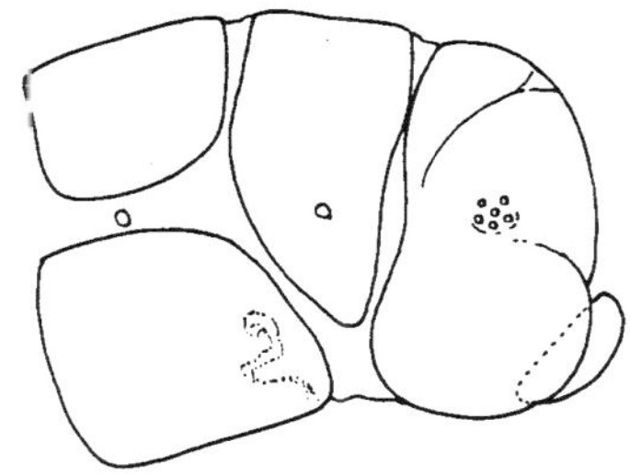

104

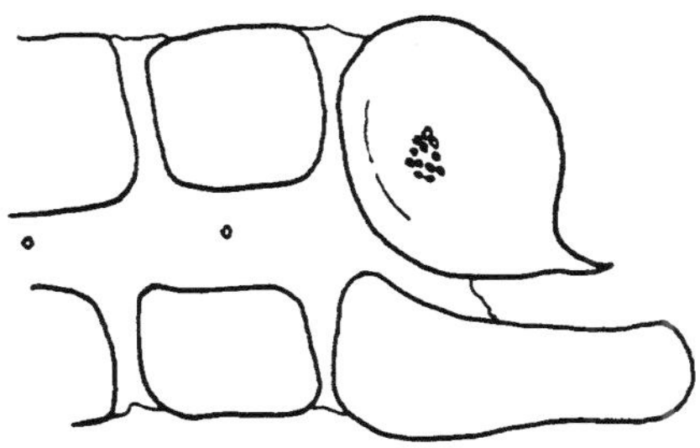

105

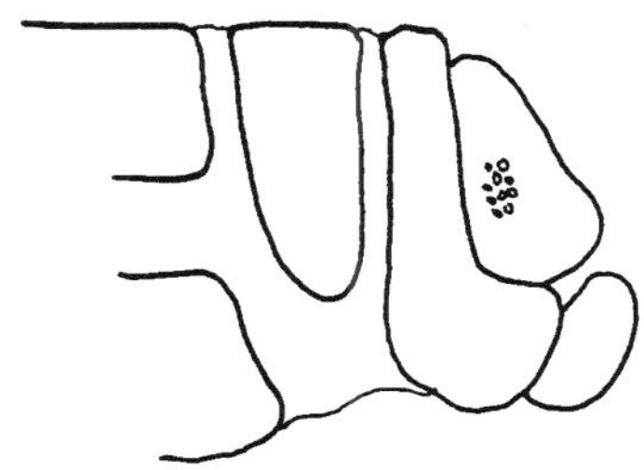

common on coffee in Papua New Guinea (New, 1989b). *Zachobiella* is perhaps the most characteristic genus of Hemerobiidae in the region, and one of very few largely endemic to it.

Megalomina Banks (type species *Megalomina acuminata* Banks; Microminae) is a small genus with only three species in Australia and Papua New Guinea. The forewings are prominently pointed. It is also separable from *Micromus* by having the humeral costal crossvein recurrent rather than simple. Synonyms in the Australian fauna are *Drepanomina* Tillyard and *Oxybiella* Tillyard.

Noius Navás (type species *Noius oceanicus* Navás; Microminae) contains a few species from islands in the western Pacific region, and has recently been recorded from New Guinea (Oswald, 1993a). It is very similar to *Micromus* in general appearance, and differs from it by relatively small forewing characters as noted in the key: most obviously, a second Sc-R crossvein is present, though this is developed only weakly in some species (it is always absent in *Micromus*), and crossveins 1M-Cu and 2M-Cu are close, separated by approximately their combined lengths (in *Micromus* 2M-Cu is absent or very short and the veins more distantly separated).

Micromus Rambur (type species *Hemerobius variegatus* Fabricius; Microminae; figs. 89, 91, 104, 105) is one of the most diverse hemerobiid genera, with about 100 nominal described species. It is the genus most likely to be encountered easily in Malesia. *M. timidus* is one of the most abundant hemerobiids in the area, and occurs on many kinds of vegetation. Many generic synonyms, or putative subgenera, have been applied, and some are relevant to the regional fauna; most were appraised and synonymised by Tjeder (1961). Those relevant to Malesian literature include *Eumicromus* Nakahara, *Archaeomicromus* Krüger, *Indomicromus* Krüger, *Stenomicromus* Krüger, *Tanca* Navás, *Idiomicromus* Nakahara, *Austromicromus* Nakahara and *Mixomicromus* Ghosh.

Notes on genera on the fringe of Malesia

Carobius Banks (type species *Carobius pulchellus* Banks; fig. 86) is an isolated genus and the sole member of the Carobiinae, regarded by Oswald (1993a) as a sister group of the remaining Hemerobiidae. It is known only from Australia (with nine species) and, although predominantly in the southern half of the continent, could possibly be found in New Guinea. Its discovery outside Australia would be of considerable interest.

Wesmaelius Krüger (type species *Hemerobius concinnus* Stephens; Hemerobiinae; including species classified as *Kimminsia* Killington; fig. 95) is one of the larger genera of the family, with nearly 70 species (Oswald, 1993a). It is predominantly a northern hemisphere temperate region genus, but some species have been dispersed elsewhere for biological control. It occurs in the eastern Palaearctic (Japan) and might occur on the northern fringes of Malesia. The identity of a species recorded from the Philippines by Navás has not been confirmed. *Kimminsia* is usually considered a synonym of *Wesmaelius*, but Oswald (1993a) found this difficult to justify. See *Hemerobius*, above, for more diagnostic detail.

Megalomus Rambur (type species *Megalomus tortricoides* Rambur), the only genus of Megalominae, includes about 40 species from many parts of the world, particularly the Neotropics and northern temperate regions. Species reported from Asia as *Megalomus* or the synonym *Boriomyia* Banks may be generically misplaced, and investigation is needed of the true status of this genus in the region. Other synonyms of *Megalomus* are *Pleomegalomus* Krüger, *Pirionus* Navás, *Allotomyia* Banks and *Spinomegalomus* Nakahara.

Neuronema McLachlan (type species *Hemerobius decisus* Walker; Drepanopteryginae) is a poorly characterised genus recorded from China, Taiwan and Japan to

India and Nepal, but probably does not occur in tropical Malesia. There are approximately 30 nominal species in the genus. Generic synonyms include *Ninguta* Navás, *Ninga* Navás, *Kulinga* Navás and *Sinoneuronema* Yang.

Drepanacra Tillyard (type species *Drepanepteryx humilis* McLachlan; Drepanacrinae; fig. 90) is one of the most characteristic hemerobiid genera in Australia, with species also in China, Taiwan, New Caledonia and New Zealand, as well as some other islands in the western Pacific. Generic synonyms include *Menopteryx* Krüger and *Drepanacrella* Kimmins.

Sympherobius Banks (type species *Hemerobius amiculus* Fitch) is the largest genus of Sympherobiinae and the only one to occur outside South America. Though reported from temperate regions of Asia it does not occur in Australasia and is probably absent also from the Old World tropics. Sympherobiinae are very distinct, and separable from most other Hemerobiidae by having only two main radial sector branches in the forewing, a feature which led historically to them being allocated a distinct family status. However, a few species of *Sympherobius* have three Rs branches.

Biological outline

Hemerobiidae are almost always associated with vegetation and are collected commonly by beating and sweeping. Considerable habitat specificity is known elsewhere, with some northern hemisphere species separated on either low vegetation or trees and, for the latter, a strong 'specificity' for either conifers or broadleaved trees, with rarer more specific limitations. As with Coniopterygidae and Chrysopidae, it is not clear whether such habitat specificity is mediated by the plant, or by 'preferences' for prey species associated clearly with one or other category of plant. Eggs are laid flat on the substrate (and are thereby differentiated from those of most Chrysopidae). Larvae are campodeiform and active, but never become coated with debris. They lack long dorsal setae.

MAJOR REFERENCES — Nakahara (1960): major synopsis of genera, now outdated, but useful as a source of history of classification and generic features used by earlier workers; Tjeder (1961): South African fauna, with summary of most earlier work on family; Oswald (1993a): world review and critical analysis of genera and relationships within the family.

Chrysopidae (green lacewings)

106

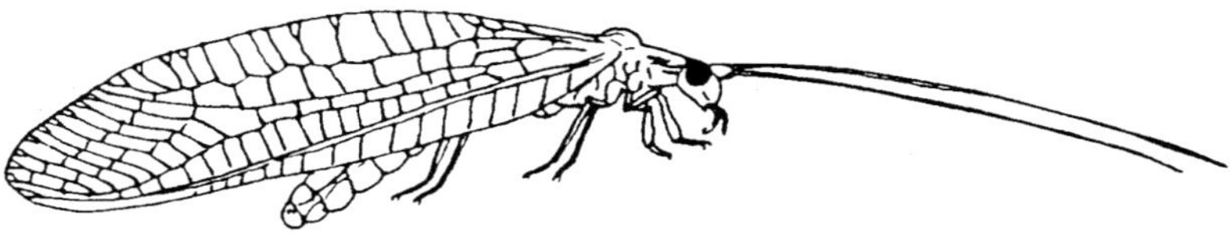

107

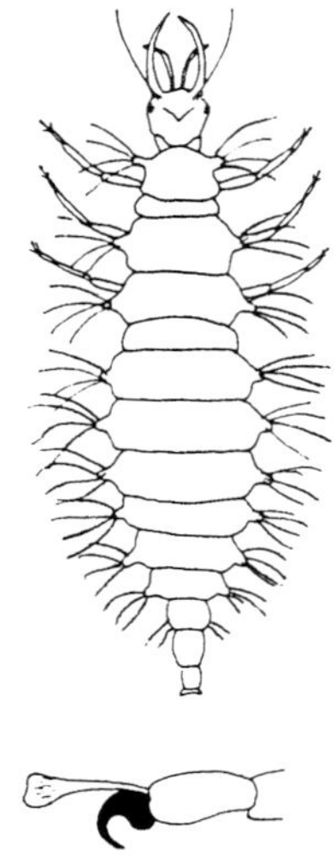

FIGURES 106 - 107
106 Adult Chrysopidae
107 larval Chrysopidae, with trumpet-shaped empodium between tarsal claws.

Family diagnosis

Slender, small to large lacewings (forewing length approximately 6-35 mm). Head medium sized; eyes prominent; ocelli absent; antennae filiform, length from approximately $^{1}/_{2}$ to >2 times length of forewing, flagellar segments with whorls of short setae; mandibles strong, pointed, symmetrical or asymmetrical; maxillary palpi five-segmented; labial palpi three-segmented. Pronotum well defined, shape variable; pterothorax broad. Legs slender, cursorial, tibiae sometimes with short spurs; tarsal claws simple or expanded basally. Wings large, subequal, narrow to broad, apex usually rounded, rarely acute; hindwing sometimes narrower than forewing; pterostigma present; trichosors absent; membrane hyaline or spotted or suffused with grey or brown. Forewing: costal cell broad, humeral crossvein simple; numerous costal crossveins, usually simple, more rarely some forked or interlinked; Sc and R1 not fused (except, rarely, in hindwing behind pterostigma), usually linked by several distal crossveins; a single basal Sc-R crossvein (absent in Apochrysinae); several branches of Rs, linked by few (most commonly two) series of gradate crossveins, gradates rarely absent or strongly reduced; intramedian cell (im) present in many taxa; branches of M beyond im fused to form 'pseudomedia'; 'pseudocubitus' formed by fusion of MA with branches of M; three anal veins present. Abdomen slender, tergites and sternites well defined.
Female: tergite ix usually expanded ventrally and fused with ectoproct; sternite vii sometimes ornamented or excised posteriorly; lateral gonapophyses large;

subgenitale present; spermatheca simple, duct slender.
Male: tergite ix distinct (some Nothochrysinae, not in Malesia) or fused with ectoproct, often with strengthening apodeme; sternite viii simple or (most Chrysopinae) fused with sternite ix to constitute compound sternite viii+ix, often with lateral apodeme. Male genitalia complex and variable: gonarcus transverse or arched; tignum sometimes present above gonarcus; entoprocessus, arcessus, pseudopenis, gonapsis each present or absent in various taxa; occasionally with parameres (*Italochrysa*); hypandrium internum small.

Diversity and distribution

Chrysopidae, the green lacewings or 'golden eyes', are among the most familiar Neuroptera to many people as a cosmopolitan family (about 1250 described species, with 105 in Malesia) with numerous very similar taxa and the general bright colours associated with their common names. Until recently, generic limitations in much of the family were very confused, but the major revision by Brooks & Barnard (1990) has done much to remedy this. As Brooks (1997) commented, the progress made has alleviated many, probably most, of the dilemmas over generic definition in Chrysopidae and genera are now based on apomorphic characters, defined on a wide range of characters comparable throughout the family, female characters are adequately defined (so that the earlier situation where many females could not be allocated confidently to genus no longer pertains), a phylogenetic framework outlined, and the number of poorly defined monotypic genera reduced considerably. Chrysopidae is the most speciose lacewing family in Malesia. However, many species remain to be described, and some described taxa have been characterised very poorly, so that their correct generic position can not be settled at present.
Although brightly coloured in life many Chrysopidae fade rapidly to brown when dry and become discoloured in alcohol. Wherever possible, notes on the appearance of fresh specimens should be made. Intraspecific and geographical variation in colour pattern may occur, so that identification on venation and genitalic features is often needed. Wing features of value (fig. 108, and later figures) include the number and position of gradate veins, the form of the intramedian (im) cell, width of the costal area, and general wing shape. Genitalic terminology is also complex, and that used here (following Tjeder, 1966) is explained in figs. 109-111.

FIGURES 108 - 111

Chrysopidae - **108** forewing of generalised chrysopine to indicate terminology — **109** generalised terminology for apical abdominal region of female, **110** of male, lateral aspect, with gonosaccus everted to display structures which are normally more cryptic), and **111** of male, posterior aspect, main structures only (c1, c2, dcc = first, second and distal cubital cells; im = intramedian cell; Psc = pseudocubitus; Psm = pseudomedia).

108

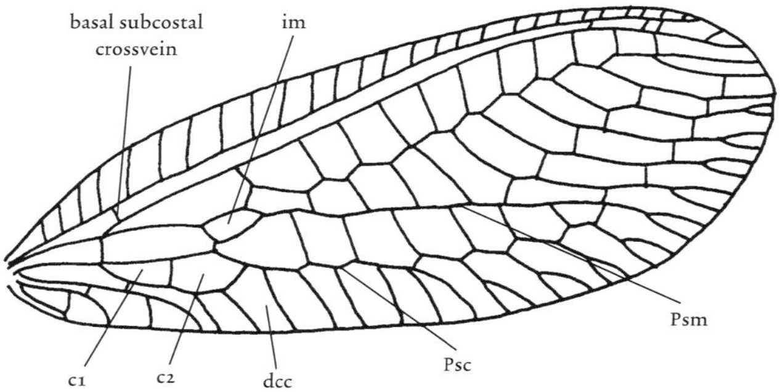

109

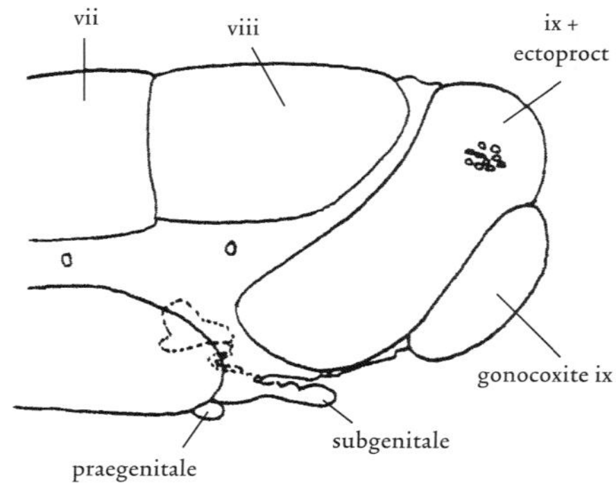

110

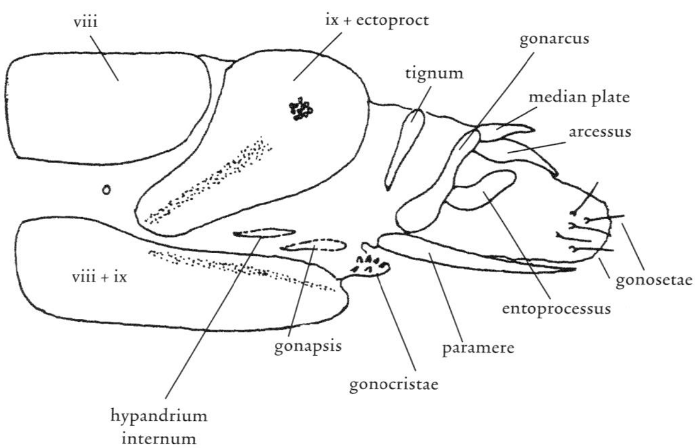

111

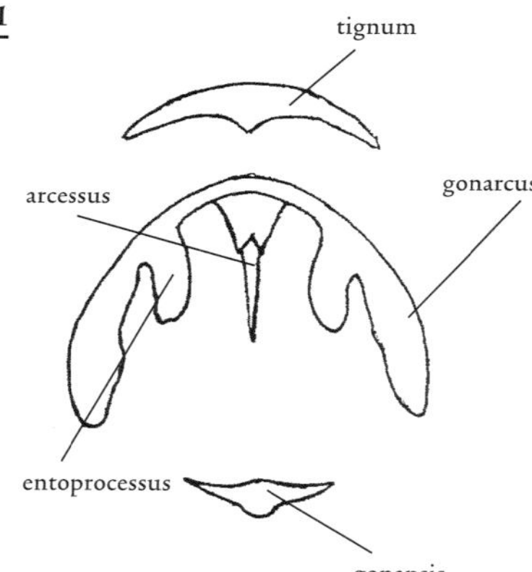

FIGURES 112 - 115

Chrysopidae - **112** forewing venation of Nothochrysinae (not in Malesia, the Australian *Triplochrysa* shown) — forewing venation of Apochrysinae: **113** *Joguina*, **114** *Nobilinus*, **115** *Synthochrysa*.

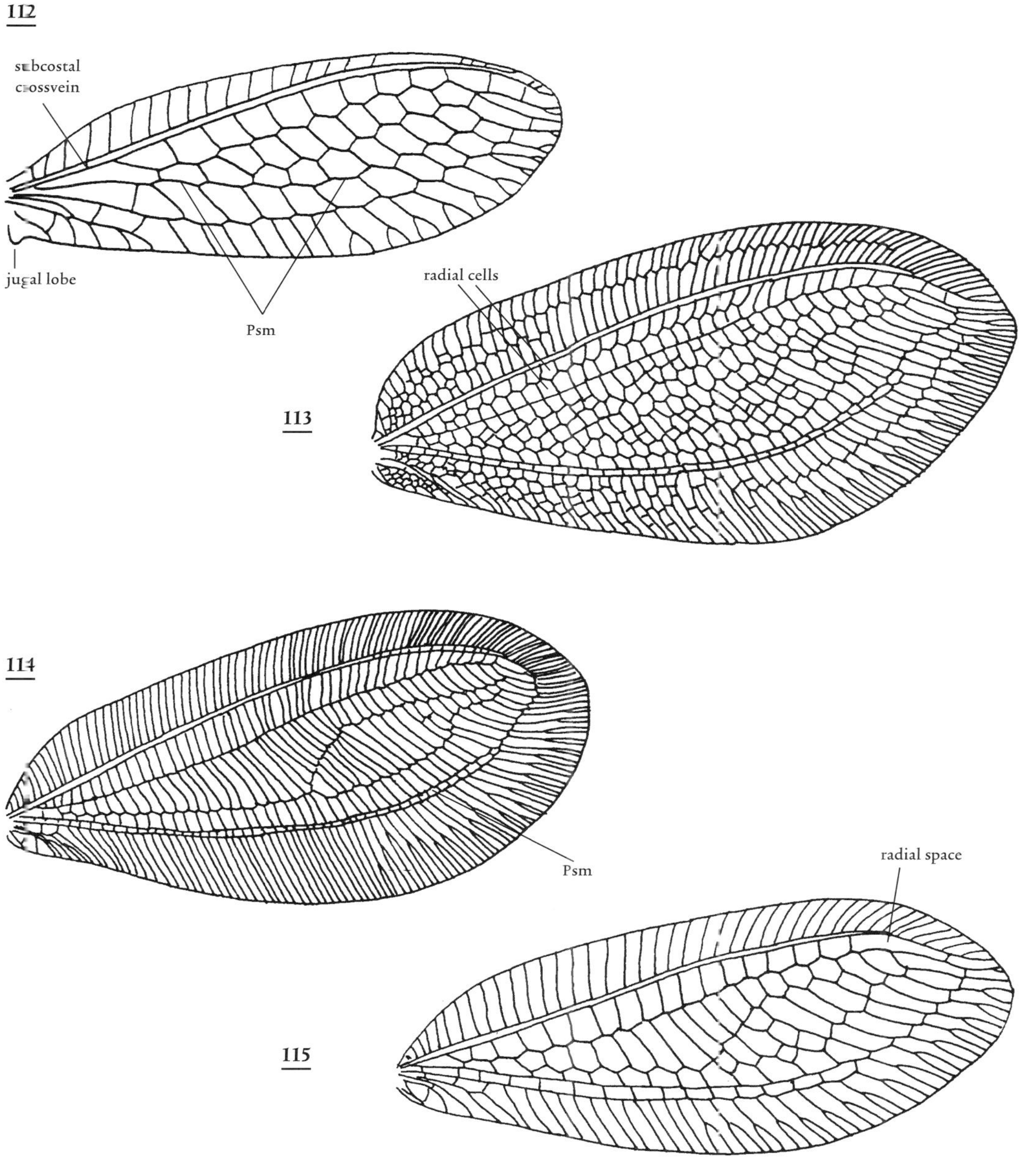

Key to the subfamilies of Chrysopinae

NOTE

Chrysopidae is divisable into three main subfamilies, with the position of one genus, the Australian *Nothancyla* Navás, still anomalous. It is currently placed in Apochrysinae, but *Nothancyla* has an im cell. The subfamilies are separated on wing venation, as follows.

1 Forewing vein Psm continuous with inner row of gradate veins; jugal lobe well developed (fig. 112) ▸ **Nothochrysinae**

– Forewing vein Psm meets outer row of gradate veins; jugal lobe small or absent ▸ 2

2 Forewing basal subcostal crossvein present; cell im present; setae on flagellar segments in four (or fewer) whorls ▸ **Chrysopinae**

– Forewing basal subcostal crossvein absent; cell im absent; setae on flagellar segments in five whorls ▸ **Apochrysinae**

NOTHOCHRYSINAE

Nothochrysinae are absent from Malesia, but two genera (*Triplochrysa* Kimmins, *Dictyochrysa* Esben-Petersen) are found in forested environments in eastern Australia. The two other subfamilies are widely distributed in Malesia and elsewhere, but many species in the region (especially in Chrysopinae) have been characterised poorly or await description. In the large subfamily Chrysopinae generic confusion has been rampant in the past, with many taxa allocated uncritically to *Chrysopa* s.l. as a 'holding genus': application of this name thus requires special care, and most described species were reallocated by Brooks & Barnard (1990). A residue of unallocated species (see checklist) attests to the remaining problems in the area.

APOCHRYSINAE

Apochrysinae includes the largest and most spectacular Chrysopidae, with three genera represented in Malesia. Although now accepted generally as distinct, it has sometimes been treated as a tribe of Chrysopinae. Nearly all the 13 genera in the subfamily have very restricted geographical ranges, and the Malesian taxa are among the most widely distributed members of the subfamily. Brooks & Barnard (1990) considered the genera of the Apochrysinae likely to be closely related to each other because of the large number of shared characters, but retained the various nominal genera as distinct, even if monotypic, because synonymy would be highly uncertain.

Key to the genera of Apochrysinae

All apochrysines are large lacewings with slender elongate bodies. They are found mainly in forest environments, and the genera seem to be distinctive. All are rather rare and taken more by chance than design. They have been collected at light and by beating or sweeping vegetation, but very little is known of their biology.

1 Radial area of forewings with at least the basal half including two or more rows of cells (fig. 113) ▸ *Joguina* Navás
– Radial area of forewings with single row of cells, although an occasional divided cell may be present ▸ 2

2 Radial area of forewings with crossveins almost to apex (wing venation very dense; fig. 114) ▸ *Nobilinus* Navás
– Radial area of forewings with distinct space free of crossveins beyond the pterostigma (fig. 115) ▸ *Synthochrysa* Needham

Notes on the genera

Joguina Navás (type species *Apochrysa nicobarica* Brauer) has been reported from Malaysia and Indonesia. It is characterised primarily by the complex venation with multiple series of gradate veins. In life the species are sometimes marked with brown or red, including a reddish median mesonotal stripe.

Nobilinus Navás (type species *Nobilinus insignitus* Navás) is an entirely Oriental genus with most described species from Indonesia. The three gradate series of crossveins, the radial area behind the pterostigma with numerous crossveins, and the forking of both veins 1A and 2A are characteristic apomorphies and, with forewing length up to around 30 mm the species are the largest pale green chrysopids in Malesia. Unlike *Joguina*, the thorax of *Nobilinus* is usually unmarked.

Synthochrysa Needham (type species *Hemerobius stigma* Girard) occurs in Indonesia and elsewhere in the western Pacific. The species are generally smaller than those of *Nobilinus* (forewing length to about 25 mm), the pronotum usually has a reddish lateral margin, and the radial area of the forewing lacks crossveins behind the pterostigma.

CHRYSOPINAE

Chrysopinae is by far the most diverse and complex subfamily of green lacewings, and contains more than 95% of species in the family. Many of the genera are separable only by detailed examination of mouthparts and genitalic structures, in some cases used in conjunction with features of wing venation. Indeed, the first two character sets are needed even to key taxa formally to major tribes, so that I elect here to follow the more practical approach of Brooks & Barnard (1990) of providing a key directly to genera without the complications of formal tribal allocation.
Three of the four tribes occur in Malesia, the absence being the New World Leucochrysini, and the listing of genera acknowledges these higher groups. Nearly 20 genera, some of them very poorly known, have been recorded from Malesia.

Little specific information is available on Malesian taxa of Chrysopinae, but recently-demonstrated taxonomic complexity elsewhere in the world (as in *Chrysoperla*) emphasises the need for very careful evaluation if correct identification is needed. Especially if species are being considered for biological control programmes, inaccurate identification may have far-reaching consequences for pest management and the success of projects involving either augmentation of native predators or classical biological control. Even closely-related species may differ considerably in their prey preferences or other biological characteristics.

Key to the genera of Chrysopinae

NOTE
Several genera have multiple entries to reflect their variability, and that the two sexes are keyed separately for many genera.

1 Vein Sc very short, with one or two crossveins below the pterostigma ▸ *Ankylopteryx* 2
– Vein Sc long; at least three or four crossveins below the pterostigma ▸ 3

2 Cell im present in forewings (fig. 116) ▸ *Ankylopteryx* subg. *Ankylopteryx*
– Cell im absent in forewings (fig. 117) ▸ *Ankylopteryx* subg. *Sencera*

FIGURES 116 - 118

Chrysopidae - Chrysopinae: wing venation of **116** *Ankylopteryx*, forewing, **117** *Sencera*, fore- and hindwings; note im cell in forewing absent, **118** *Plesiochrysa*, forewing.

116

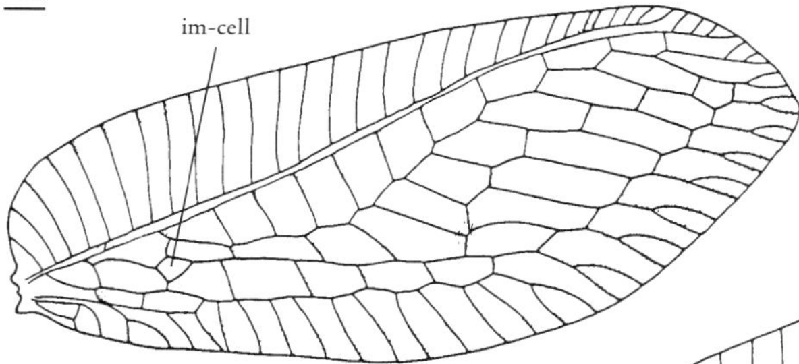

117

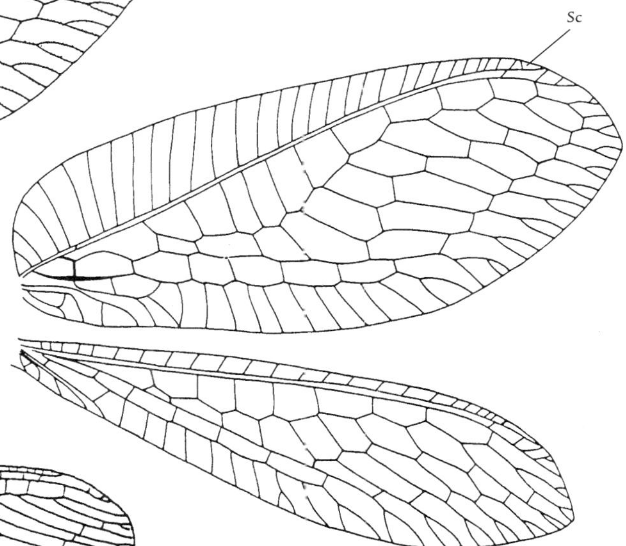

118

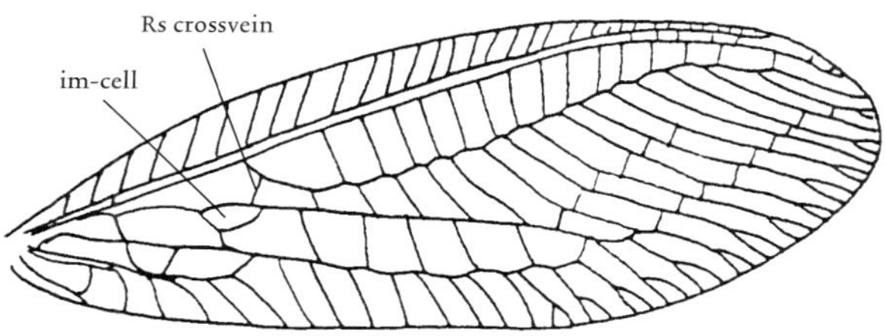

3	Forewings with three or more series of gradate crossveins ▸	4
–	Forewings with two series of gradate crossveins (occasional ones may be doubled) ▸	9
4	Hindwings with two series of gradate crossveins (antennae shorter than forewing; forewing unmarked) ▸	*Borniochrysa*
–	Hindwings with more than two series of gradate crossveins ▸	5
5	Cell im quadrangular (fig. 122) ▸	*Austrochrysa*
–	Cell im ovate (fig. 118) ▸	6

6 Costal setae long, erect or slightly inclined toward forewing apex; male sternite viii+ix elongate, extending beyond apex of ectoprocts ▸ *Chrysopidia*

– Costal setae short, strongly inclined toward forewing apex; male sternite viii+ix short ▸ 7

7 Larger species (forewing length >16 mm); male sternites viii and ix not fused, pseudopenis present ▸ *Plesiochrysa*

– Smaller species (forewing length <15 mm); male sternites viii and ix fused, pseudopenis present or absent ▸ 8

8 Female: ventral impression to spermatheca (fig. 119) sometimes absent; male with inner projection at apex of sternites viii+ix, apex of ectoprocts angled or tapered, genitalic complex variously reduced ▸ *Mallada*

– Female: ventral impression to spermatheca always present (fig. 120); male without projection at apex of sternites viii + ix, ectoprocts broad and rounded, genitalic complex complete ▸ *Dichochrysa*

9 Cell im quadrangular ▸ 10

– Cell im ovate or triangular ▸ 14

10 Pterostigma of forewings marked with dark spot (antennae equal to or shorter than length of forewing) ▸ *Stigmachrysa*

– Pterostigma of forewings unmarked or, at most, with small dark spot on base ▸ 11

11 Flagellar segments twice as broad as long; cell c1 shorter than c2; pronotum narrow ▸ 12

– Flagellar segments about as broad as long; cell c1 at least 1.5 times longer than c2; pronotum broad ▸ 13

12 Forewings with large spots on discal area (fig. 121); inner gradate crossvein series irregular, diverging widely from outer gradate series; outer gradates green ▸ *Kostka*

– Forewings without large spots; inner and outer gradate crossvein series parallel (fig. 122); outer gradates black ▸ *Austrochrysa*

13 Cell im short; male: parameres absent; female: praegenitale present ▸ *Evanochrysa*

– Cell im long (fig. 123); male (fig. 125): parameres present; female: praegenitale absent ▸ *Italochrysa*

FIGURES 119 - 123

Chrysopidae - Chrysopinae: spermatheca **119** with or **120** without ventral impression — wing venation of **121** *Kostka*, forewing, **122** *Austrochrysa*, forewing ([NOTE] im cell quadrangular compared with oval im cell of most other genera), **123** *Italochrysa*, fore- and hindwings; [NOTE] large quadrangular im cell in forewing.

119

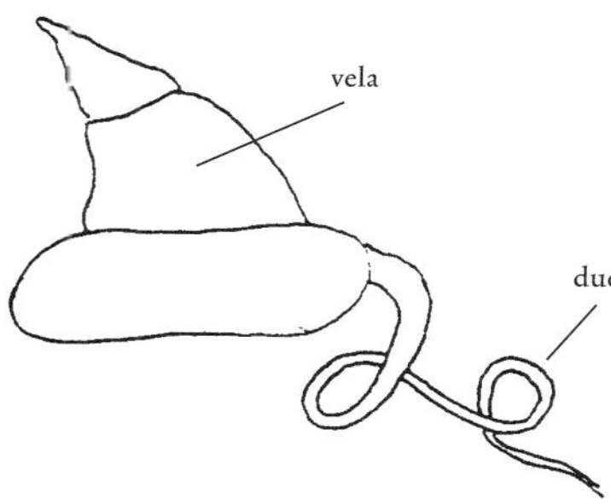

120

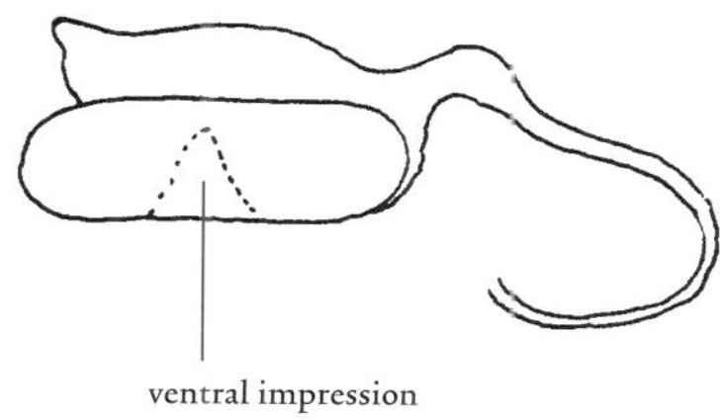

121

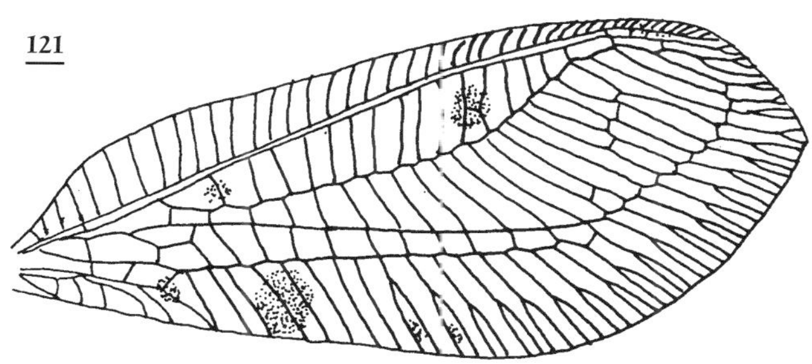

122

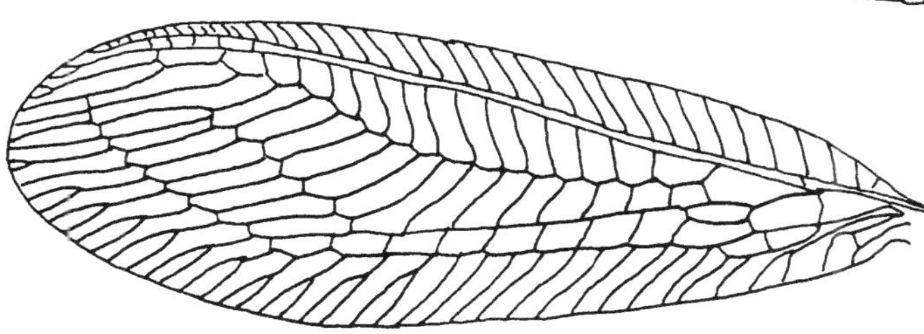

123

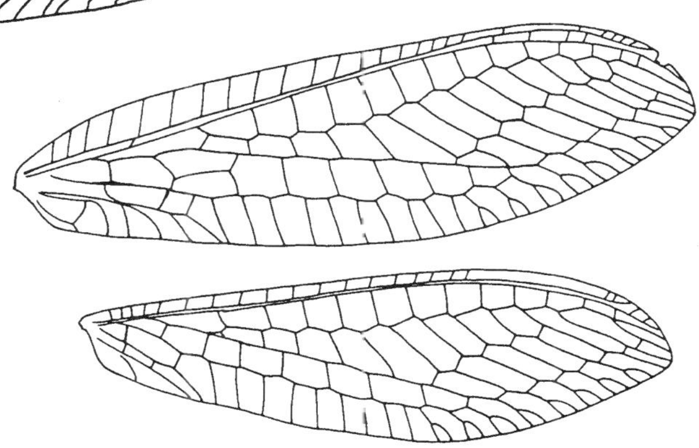

FIGURES 124 - 126
Chrysopidae - Chrysopinae: genitalic features, male abdominal apex (lateral aspect) of **124** *Italochrysa*, **125** *Mallada*, **126** *Dichochrysa*.

124

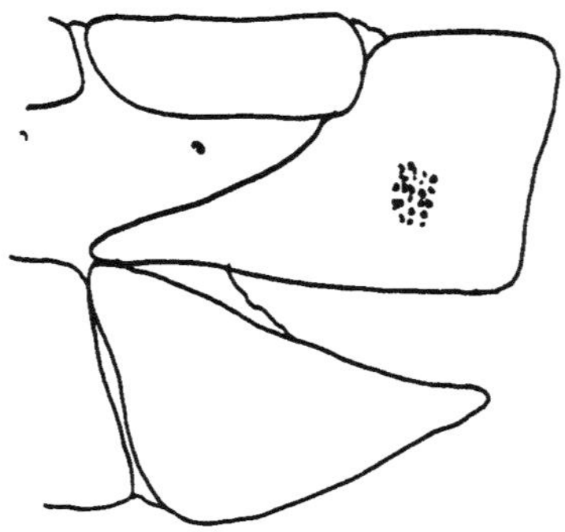

125

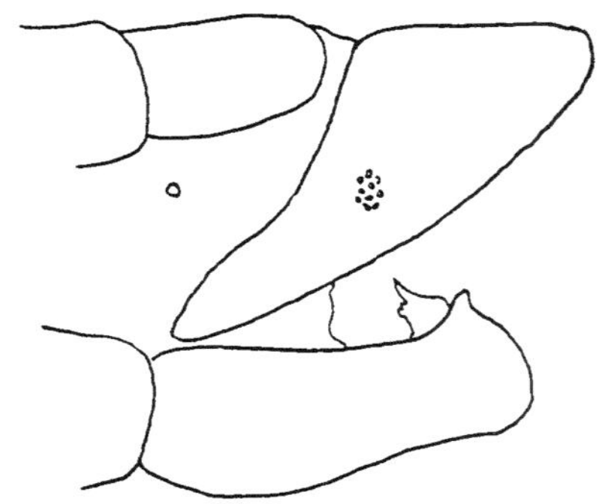

126

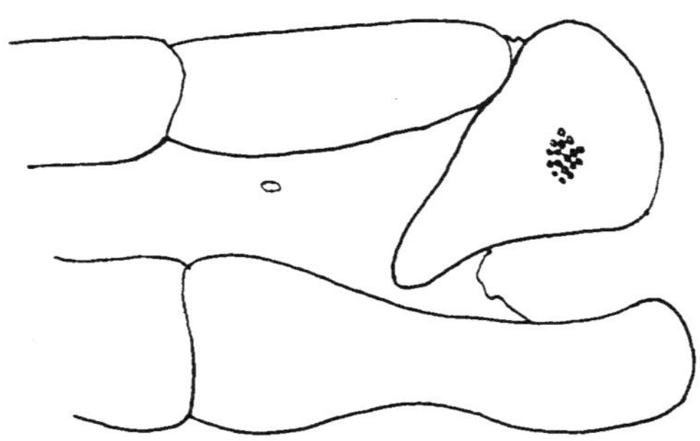

14 Cell c1 much longer than c2 ▸ ***Nodochrysa***
– Cell c1 shorter (or, rarely, very slightly longer) than c2 ▸ 15

15 Male ▸ 16
– Female (note that some parts of this key are difficult to use and some females unassociated with males provide problems for correct allocation) ▸ 28

16 Gonapsis present, though sometimes small (*Glenochrysa*, which has heavily marked wings) ▸ 17
– Gonapsis absent ▸ 21

17 Tignum present (fig. 110) ▸ 18
– Tignum absent ▸ 19

FIGURES 127 - 128
Chrysopidae - Chrysopinae: wing venation of **127** *Glenochrysa*, **128** *Mallada*.

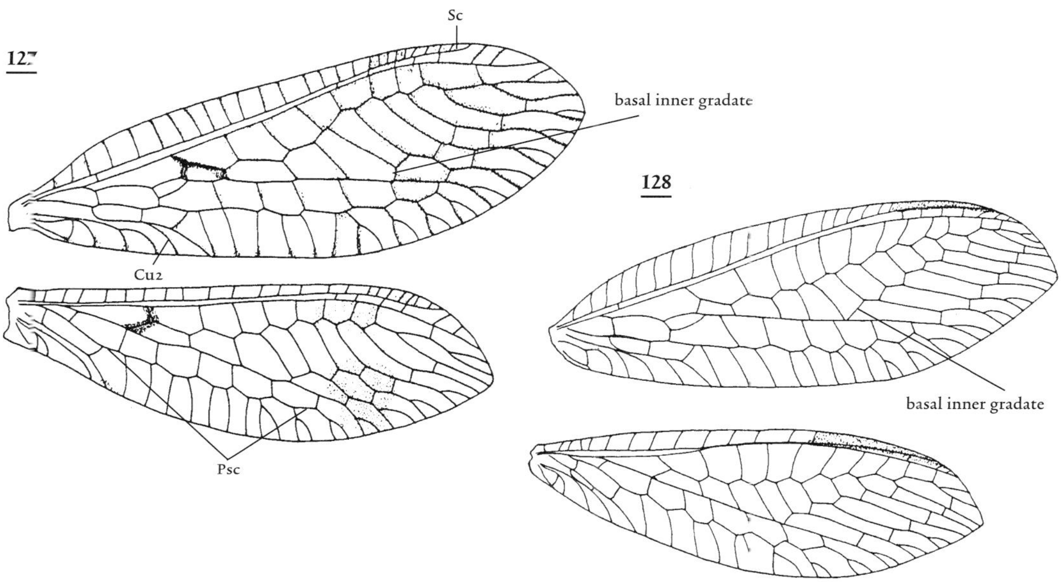

18 Sternites viii+ ix with inner projection at apex; ectoprocts usually angled and tapered at apex; genitalic complex reduced, with one or more elements absent (fig. 125) ▸ *Mallada*

– Sternites viii + ix without inner projection at apex; ectoprocts broad, rounded; genitalic complex complete (fig. 126) ▸ *Dichochrysa*

19 Forewings marked, often extensively, with black or pale brown; vein Sc rather short (fig. 127) ▸ *Glenochrysa*

– Forewings unmarked, or markings restricted to small black/brown spots on cell dcc and/or pterostigma; vein Sc long ▸ 20

20 Forewings usually with basal red suffusion; ectoprocts long and narrow, separated by dorsal suture (fig. 129) ▸ *Chrysocerca*

– Forewings unmarked; ectoprocts short, broad, fused dorsally ▸ *Apertochrysa*

21 Tignum present ▸ 22

– Tignum absent ▸ 23

FIGURES 129 - 132
Chrysopidae - Chrysopinae: genitalic features, male abdominal apex (lateral aspect) of **129** *Chrysocerca*, **130** *Chrysoperla*, **131** *Plesiochrysa*, **132** *Borniochrysa*.

129

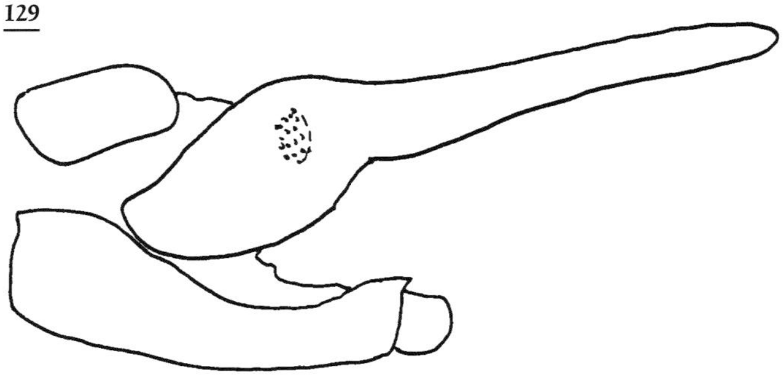

130

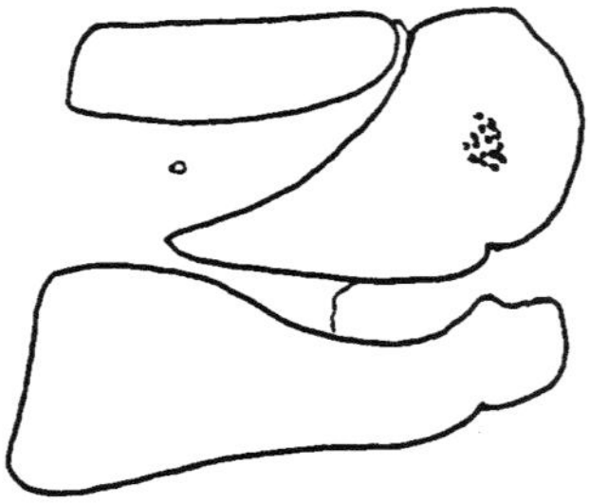

131

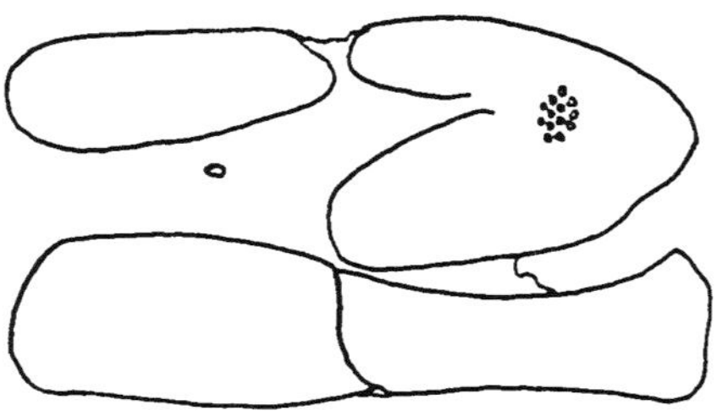

132

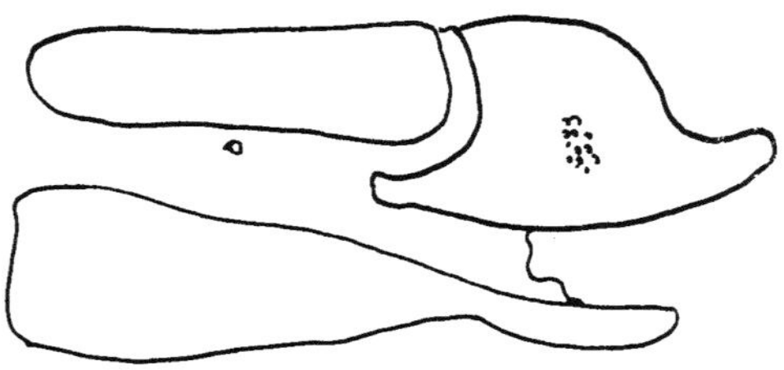

22 First Rs crossvein meets Psm distal to cell im; arcessus present; pseudopenis absent; sternites viii+ix fused; apex of sternites viii+ix with distinct 'lip' (fig. 130) ▸ *Chrysoperla*

– First Rs crossvein meets cell im before its apex (fig. 118); arcessus absent; pseudopenis present; sternites viii and ix not fused; apex of sternite ix without lip (fig. 131) (head and pronotum often with orange/red spots) ▸ *Plesiochrysa*

23 Stridulatory furrow laterally on abdominal segment 2 ▸ *Brinckochrysa* (in Malesia?)

– Stridulatory furrow absent ▸ 24

24 Sternites viii and ix not fused ▸ *Plesiochrysa*

– Sternites viii and ix fused ▸ 25

FIGURES 133 - 138
Chrysopidae - Chrysopinae: male abdominal apex of **133** *Retipenna*, **134** *Semachrysa* — character states: **135** subgenitale tapering basally (*Apertochrysa*) or **136** straight; adult mandibles **137** narrow, tapered, **138** broad.

133
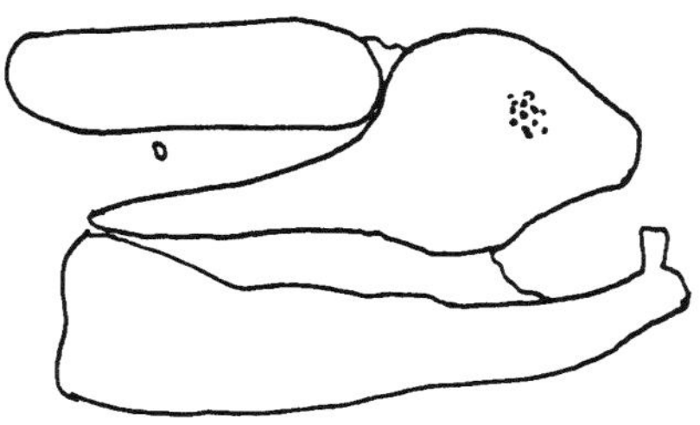

134
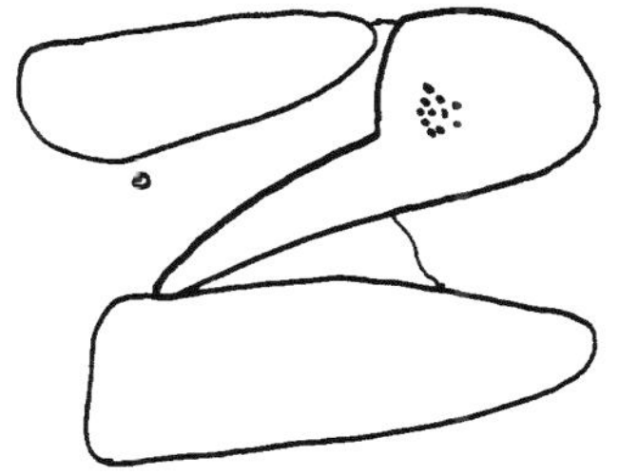

135
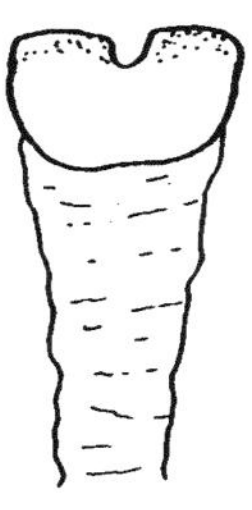

136
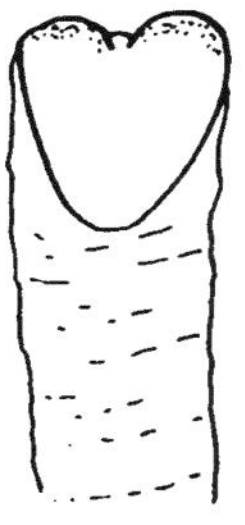

137
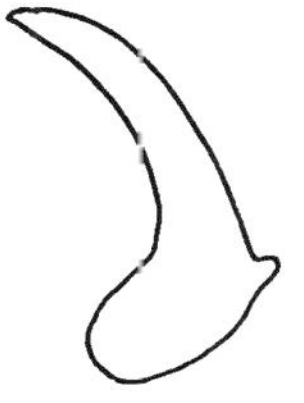

138
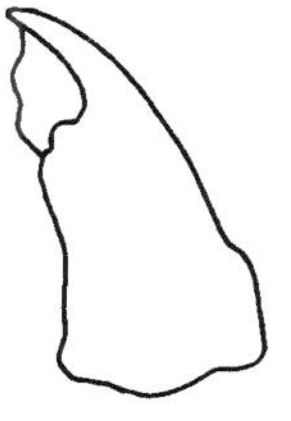

25 Ectoprocts narrow and projecting apically, with apical tuft of long setae (fig. 132) ▸ *Borniochrysa*

– Ectoprocts rounded apically, no apical tuft of long setae ▸ 26

26 Radial crossveins oblique or sinuate; scape usually elongate ▸ *Austrochrysa*

– Radial crossveins straight; scape about as broad as long ▸ 27

27 Antennae shorter than forewings; frons unmarked; apodeme of sternites viii and ix long, with prominent apical tooth (fig. 133); numerous gonosetae ▸ *Retipenna*

– Antennae longer than forewings; frons with two or three black spots; apodeme of sternites viii and ix short, not projecting (fig. 134); no more than four gonosetae on gonosaccus ▸ *Semachrysa*

28 Basal inner gradate in forewings not meeting Psm (fig. 127) ▸ 29

– Basal inner gradate in forewings meeting Psm in at least one wing (fig. 128) ▸ 34

29 Sternite vii with deep apical invagination; stridulatory furrow laterally on abdominal segment 2 ▸ **Brinckochrysa** (in Malesia?)
– Sternite vii straight or convex apically; stridulatory furrow absent ▸ 30

30 Radial crossveins sinuate or oblique ▸ *Plesiochrysa*
– Radial crossveins straight ▸ 31

31 Sc short; basal branch of vein Cu2 recurrent; wing veins usually with extensive black or pale brown shading (fig. 127) ▸ *Glenochrysa*
– Sc long; basal branch of vein Cu2 not recurrent; wings unmarked or with only small spots ▸ 32

32 Subgenitale tapering basally (fig. 135) ▸ *Apertochrysa*
– Subgenitale straight basally (fig. 136) ▸ 33

33 (Females of the two genera keyed here can not be separated reliably as the sole character given is variable; after Tsukaguchi 1995, for Japanese fauna.)
Spermatheca sometimes without ventral impression ▸ *Mallada*
– Spermatheca always with ventral impression ▸ *Dichochrysa*

34 Radial crossveins sinuate or oblique ▸ 35
– Radial crossveins straight ▸ 37

35 Median radial crossveins sinuate; radial crossveins below pterostigma straight (fig. 122) ▸ *Austrochrysa*
– Median radial crossveins straight; radial crossveins below pterostigma oblique ▸ 36

36 (Females of the two genera keyed here can not be separated reliably as the sole character given is variable; after Tsukaguchi 1995, for Japanese fauna.)
Spermatheca sometimes without ventral impression ▸ *Mallada*
– Spermatheca always with ventral impression ▸ *Dichochrysa*

37 Wings often extensively marked; palpi narrow, elongate apically, rounded in cross-section; mandibles narrow, scythelike (fig. 137) ▸ 38
– Wings unmarked; palpi broad, tapering abruptly at apex, flattened in cross-section; mandibles broad (fig. 138) ▸ 39

38 Frons unmarked; spermathecal duct with broad ventral expansion ▸ *Retipenna*
– Frons with two or three black spots; spermathecal duct narrow ▸ *Semachrysa*

FIGURES 139 - 140
Chrysopidae - Chrysopinae: forewing venation of **139** *Chrysoperla*, **140** *Chrysocerca*.

139

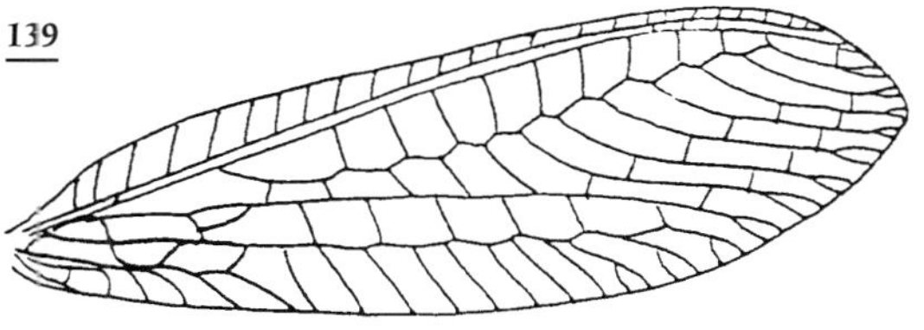

140

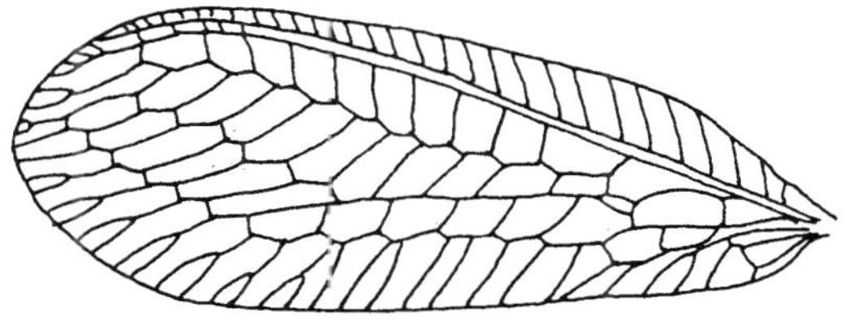

39 First Rs crossvein meets Psm distal to im cell (fig. 139) ▸ *Chrysoperla*
– First Rs crossvein meets im cell at apex or subapically (fig. 140) ▸ 40

40 Base of forewings and mesonotum marked with red/brown/black *Chrysocerca*
– Base of forewings unmarked ▸ 41

41 Basal fork of M (at base of im cell) and crossvein between M1 and M2 thickened; head (vertex and frons) usually marked with yellow/orange/red suffusion ▸ *Plesiochrysa*
– Crossveins at base of forewing not thickened; head (vertex and frons) with black or brown marking; if red, this limited to small spot or narrow stripe rather than broad suffusion ▸ *Borniochrysa*

Notes on the genera

Tribe Ankylopterygini

Ankylopteryx Brauer (type species *Chrysopa venusta* Hagen) is divided into two subgenera - *Ankylopteryx* s.str. and *Sencera* Navás - with the former the more diverse. *Ankylopteryx* s. str. contains two major species groups, both found in Malesia. One group (typified by *A. octomaculata* (Fabricius) and its forms) has large dark brown or black spots on the wings, while the other lacks such spots but has the wing veins suffused or 'shaded' with pale grey or brown. The subgenus *Sencera* (formerly treated as a separate genus) differs from *Ankylopteryx* s.str. consistently only in lacking the im cell. *Sencera* contains only four described species, but a few others are known.

Retipenna Brooks (type species *Chrysopa notata* Navás) is distinguishable from other ankylopterygine genera by the relatively broad hindwings, the widely separated veins Sc and R, and the short gonosaccus. Females are distinctive on the broad ventral expansion of the spermathecal duct (Brooks & Barnard, 1990). Eight species are known, most of them highly localised. Only one, the widely distributed *R. notatus*, has been recorded from Malesia.

Semachrysa Brooks (type species *Semachrysa minuta* Brooks) is a replacement name for *Indochrysa* Banks, which is invalid because no type species was designated. It extends from Japan, throughout Southeast Asia to northern Australia. The 14 species recognised by Brooks (1983) can be divided among three species groups on differences in male genitalia. Eight species have been reported from Malesia; males of three of these remain unknown.

The larva of the Japanese *S. matsumurae* (Okamoto) has been reported feeding on coccids.

NOTE

Also *Signochrysa* Brooks & Barnard (type species *Leucochrysa mira* Navás), to which one '*Chrysopa* s.l.' species (*C. valdezi* Banks, from Borneo) was allocated tentatively by Brooks & Barnard (1990). It is not included in the above key: species of *Signochrysa* are small to medium sized (forewing length to about 12 mm), with the forewing marked with black spots or shading, mandibles narrow and symmetrical, elongate antennal scape, and divergent radial crossveins behind the forewing pterostigma.

Tribe Belonopterygini

Evanochrysa Brooks & Barnard (type species *Nothochrysa evanescens* McLachlan), with three Indo-Malaysian species, is very similar in general appearance to *Italochrysa* and differs from it on relatively small features of wing venation, namely a shorter im cell, a short second median cell, and with the inner gradate veins basally extended. The males lack parameres, which are conspicuous in males of *Italochrysa*.

Italochrysa Principi (type species *Hemerobius italicus* Rossi) is by far the largest genus of Belonopterygini in the region, with about twelve regional species described and others known. They are also the sturdiest Chrysopidae and the largest Chrysopinae in the region (with the forewing of some species exceeding 25 mm in length). Many species have strongly contrasting black or brown markings on the body and black or brown venation. The long quadrangular im cell is characteristic. Many of the included species were earlier known as '*Nothochrysa*', and recognition of some species is rather tentative.

Nodochrysa Banks (type species *Chrysopa necrota* Banks) is known from a single species from Malaysia, Singapore and Brunei. It was for long a very poorly diagnosed genus, as it was described briefly in a key couplet and not investigated further for some time (Brooks & Barnard, 1990).

Stigmachrysa Navás (type species *Nothochrysa kervillei* Navás), another small genus extending from India to Central Indonesia, is also poorly known.

Tribe Chrysopini

Generic limits in Chrysopini are sometimes very difficult to define, and the complex nature of the male genitalia renders generic diagnoses difficult. Considerable care may be needed in examining genitalic components and the generic definitions given here in the key rely largely on these features and, inevitably, are somewhat more complex than for most other Malesian Neuroptera. However, because of the values of - or at least associations with - particular taxa for biological control of pests, precise diagnoses and identifications are important also in applied contexts, in addition to faunal evaluation. Several of the genera listed here were treated historically as synonyms of *Chrysopa* s.l. or, more lately, as subgenera following Tjeder (1966), who first described the considerable taxonomic values of the complex male genitalia. *Chrysopa* s.str. is a Holarctic genus, and all Malesian taxa previously allocated to it should properly be referred to other genera. However, as noted earlier, a residue of unallocated species remains to be revised.

Apertochrysa Tjeder (type species *Chrysopa umbrosa* Navás) contains relatively small lacewings, apparently closely related to *Mallada* and *Dichochrysa* and resembling them closely on external features. It was formerly treated as a subgenus of *Chrysopa* or *Mallada*, and raised to generic status by Tsukaguchi (1985). Undescribed species are known from Malesia, but only one described species is allocated to this genus.

Austrochrysa Esben-Petersen (type species *Austrochrysa samoana* Esben-Petersen) is predominantly Oriental, with several species recorded from Indonesia. It is unusual in the Chrysopini in having a quadrangular im cell and, in some species, a third row of gradate veins in one or both wings. *Scoliochrysa* Navás is a synonym.

FIGURES 141 - 142
Chrysopidae - Chrysopinae: female abdominal structures, abdominal apex (lateral aspect) of **141** *Brinckochrysa*, **142** *Chrysoperla*.

141

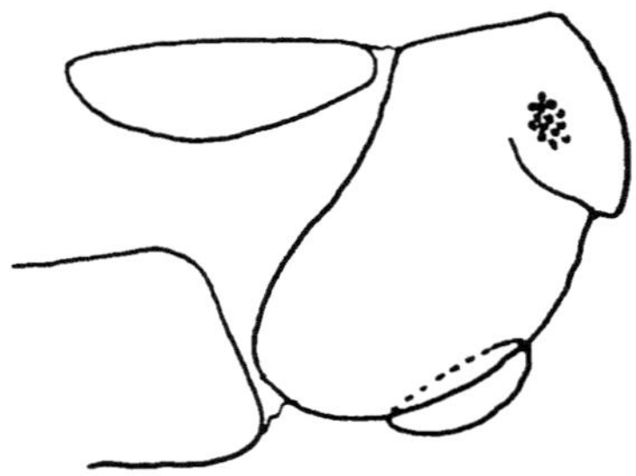

142

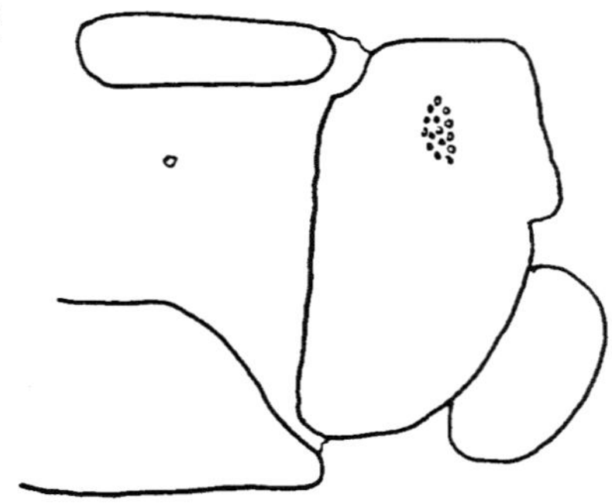

Borniochrysa Brooks & Barnard (type species *Bornia winkleri* Navás), a replacement name for the preoccupied *Bornia* Navás, also has three rows of gradate veins, but the im cell is not triangular.
Brinckochrysa Tjeder (type species *Chrysopa (Brinckochrysa) peri* Tjeder) has not been recorded from Malesia. It is widely distributed from tropical Africa to Micronesia and occurs in northern Australia. It may well occur in the region. *Brinckochrysa* is distinctive in having a stridulatory furrow on the second abdominal segment. *Neda* Navás is a synonym.

Chrysocerca van der Weele (type species *Chrysocerca jacobsoni* van der Weele) (fig. 140). Other than for a single African species, *Chrysocerca* is restricted to the Oriental region. Males are very distinctive because of their elongate ectoprocts (fig. 129), and Brooks & Barnard (1990) noted that females can be associated on the long antennae (considerably longer than the forewing), the large red and black spot on the mesonotum of some species, and the relatively large cercal callus. *Pseudochrysa* Okamoto is a synonym.

Chrysoperla Steinmann (type species *Chrysopa carnea* Stephens) is of particular interest for biological control, and was monographed by Brooks (1994). It is very widely distributed, with complex sibling species groups in Europe and the Nearctic region. *Chrysoperla* was formerly a subgenus of *Chrysopa*.

Chrysopidia Navás (type species *Chrysopidia nigrata* Navás) contains one regional species, from the Philippines. The genus is widespread in the Palaearctic and is recognised by the elongate scape; males have sternites viii + ix greatly elongated. Only one of the three constituent subgenera, *Chrysotropia* Navás, occurs in the area. This was formerly treated as a subgenus of *Chrysopa*, then a full genus, but considered a subgenus of *Chrysopidia* by Brooks & Barnard (1990), whose appraisal is followed here.

Glenochrysa Esben-Petersen (type species *Glenochrysa typica* Esben-Petersen) has few species (some undescribed) in Malesia. These are fairly small chrysopids, with a forewing length only around 10-13 mm. They have extensive black or dark brown wing markings with a characteristic brown iridescence. Parts of the wing membrane may be embossed. The appearance of most *Glenochrysa* species is very distinctive.

Kostka Navás (type species *Kostka nacaratus* Navás). This rare monotypic genus is known only from two specimens from Sumatra. *K. nacaratus* is a large lacewing with spotted forewings, numerous Rs branches and a highly irregular inner gradate series (fig. 121).

Mallada Navás (type species *Mallada stigmatus* Navás) (fig. 128) is a worldwide genus, but the differentiation of *Dichochrysa* (below) has restricted it to a relatively small number of species. Many of the species were described originally as *Anisochrysa* Nakahara, now a synonym, as is *Triadochrysa* Adams.

Dichochrysa Yang (type species *Navásius eumorphus* Yang & Yang), raised as a new name for the unavailable *Navásius* Yang & Yang, differs from *Mallada* in small but consistent features of male genitalia, and is proving to be a very large and complex genus, very widely distributed in Malesia and nearby countries. Most of the species allocated previously to *Mallada* and its synonyms are more properly placed in *Dichochrysa*, but most have not yet been transferred formally. Many females of the two genera can not yet be differentiated reliably. The major difference between the two genera is the form of the male genitalia. Those of *Mallada* are reduced in various ways, whilst those of *Dichochrysa* are complete, with tignum, gonapsis, entoprocessus and pseudopenis. The male sternites viii+ix lack a projection at the apex, and the ectoprocts are broad and rounded. In *Mallada*, males have a lobe or projection at the apex of the composite sternite, and the ectoprocts are narrowed and often tapered apically. *Pseudomallada* Tsukaguchi is a synonym.

Plesiochrysa Adams (*type species Chrysopa brasiliensis* Schneider) contains a group of large chrysopids, many of them with the pronotum elongate and with long narrow wings. Brooks & Barnard (1990) recognised four distinct species groups in Plesiochrysa, one Neotropical and the others from the Oriental/Australasian regions. *P. ramburi* (Schneider), with very characteristic black markings on the head and thorax, is very widely distributed in the western Pacific. Most other species have large red or yellow spots on the head, but one species from the Solomons has the head unmarked.

Biological outline

Some Chrysopidae, especially members of Chrysopinae, are of considerable importance in biological control, particularly as manipulable generalist predators of a variety of small arthropods on field and orchard crops. Most Chrysopinae are therefore associated with vegetation. Adults of many species are attracted to lights, but some are also susceptible to attractant baits such as methyl eugenol.

Particular species may be associated with particular vegetation formations (such as trees or low vegetation) or plant species. Eggs of most species are laid on stalks from 2-5 mm in length. Fecundity is commonly in the order of several hundred eggs. Larvae are active and campodeiform. Those of some species become covered with debris: for example, remains of prey organisms can be entangled in long dorsal hairs on the thorax and abdomen and are believed to provide the larva with effective camouflage during foraging. Adults of many species are predators, but others feed on honeydew and pollen. These species often harbour symbiotic yeasts (*Torulopsis*) thought to aid in digestion of honeydew.

Courtship may involve the use of substrate-borne 'calls' which differ clearly between closely related species in morphologically confusing sibling species groups, such as the Holarctic *Chrysoperla carnea* group. Much recent information on Chrysopidae is summarised in Canard et al. (1984) and McEwen et al. (2001).

MAJOR REFERENCES — Tjeder (1966): South African fauna, with much information on generic limits and summaries of earlier work on family; Canard et al. (1984): biological information on the family; Brooks & Barnard (1990): world synopsis of genera, the major source of information on generic and higher classification.

Nymphidae (no established vernacular name)

143

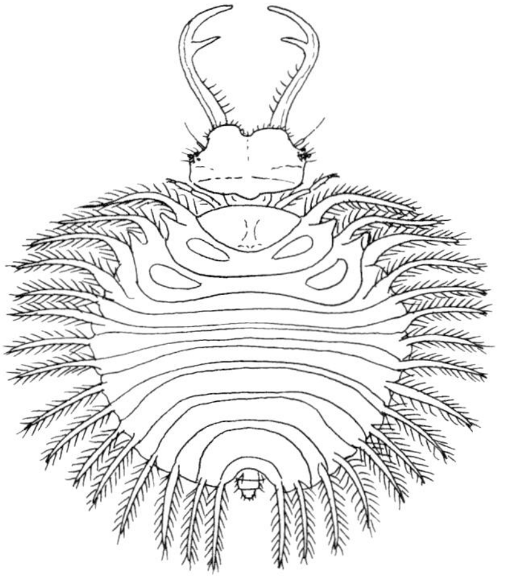

FIGURE 143
Larval Nymphidae

Family diagnosis

Medium to large lacewings (forewing length approximately 18 to over 40 mm); often bright green in life and so look like large chrysopids. Head small to medium; eyes large; ocelli absent; antennae filiform, usually 1/3-1/2 of the forewing length. Thorax slender, legs cursorial, with or without apical tibial spurs. Wings slender to very broad, subequal; trichosors present; pterostigma well-defined. Costal space slender to moderately broad, numerous costal crossveins, often forked, more rarely interlinked, humeral crossvein simple. Sc and R1 fused apically; one to many subcostal crossveins; Rs pectinate with numerous irregular gradate veins; MA arises from Rs in forewing, more distinct basally in hindwing; position of MP fork variable; CuA (forewing) or MP (hindwing) forked widely to enclose much of posterior margin, numerous crossveins between branches; three anal veins in forewing. Abdomen slender; ectoproct simple, cercal callus distinct; tergite ix often partially subtending ectoproct.

Female (fig. 144): tergite viii deep; sternite vii shallow; lateral gonapophyses convex, unornamented; subgenitale distinct; spermatheca slender.
Male (fig. 145): sternite ix simple, rounded; tergite ix sometimes produced ventrally; gonarcus present; entoprocessus usually present; arcessus present, may be ornamented ventrally; parameres present, gonapsis present or absent; hypandrium internum small.

FIGURES 144 - 148
Nymphidae - general appearance of abdominal apex, lateral aspect, of **144** female and **145** male nymphid — **146** wing venation of *Osmylops*, **147** of *Myiodactylus*, **148** of basal forewing costal region of *Umbranymphes* to show divided costal cells.

144

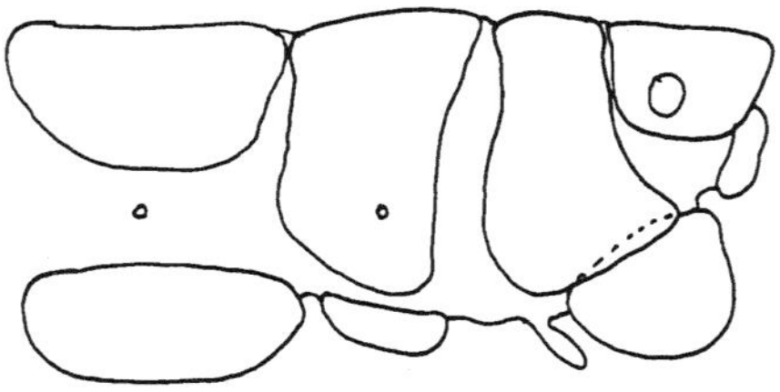

145

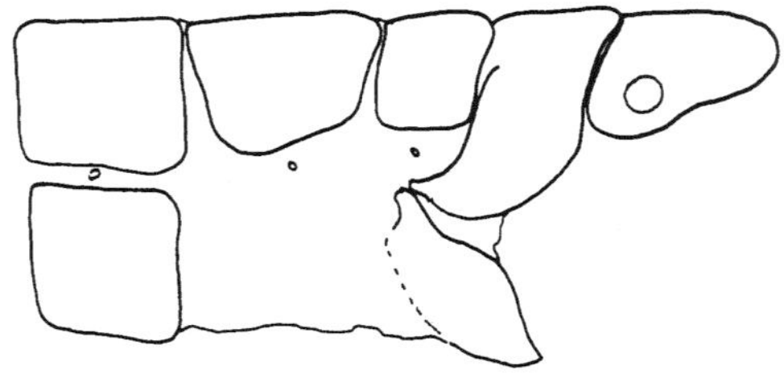

146

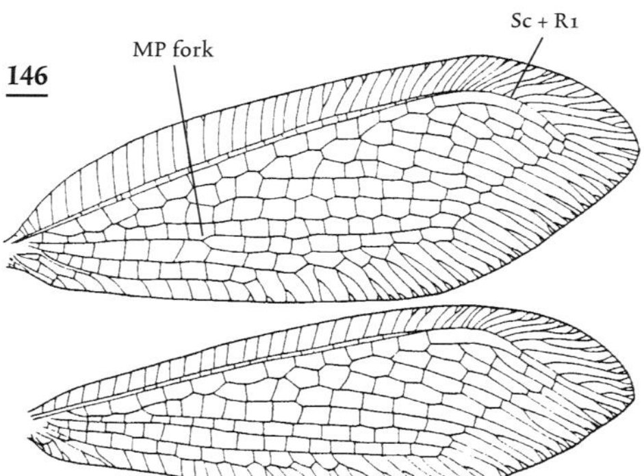

147

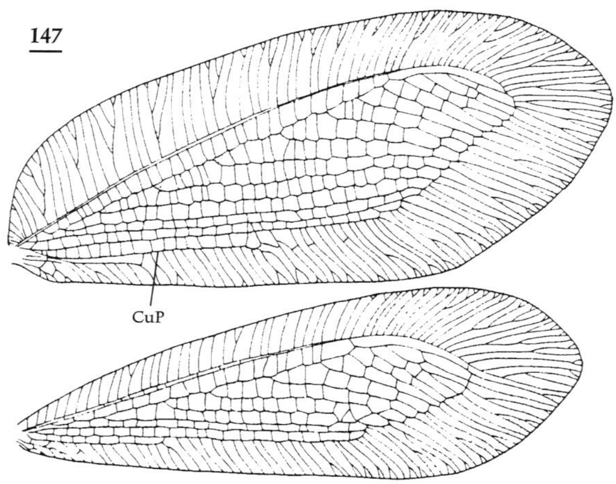

148

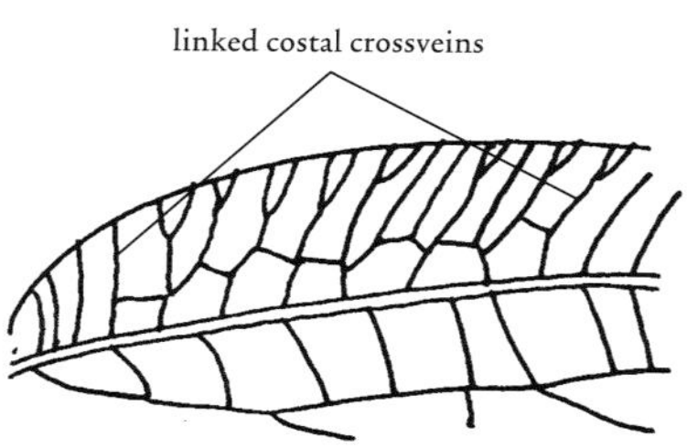

Diversity and distribution

The Nymphidae are a small (35 species) complex family found in Australia and New Guinea, with a single record from the Philippines. Nine species are known from Malesia. The family appears to be most diverse in Australia. The extra-Australian taxa are closely related and comprise members of the former family Myiodactylidae, one of two historical family segregates of Nymphidae. The former Myiodactylidae are separable from other Nymphidae by lacking tibial spurs. Three genera are known outside Australia. One of these (*Umbranymphes*) is known only from New Guinea, the other two are present also in Australia and in the Philippines (only one specimen).

Key to the genera of Nymphidae

1 Tibial spurs present (fig. 6) ▸ extralimital genera: check key in New (1981)
– Tibial spurs absent ▸ 2

2 Forewing vein MP forking in basal half of wings, usually before separation of first branch of Rs from Rs stem (fig. 146) ▸ ***Osmylops***
– Forewing vein MP apparently simple, or forking in apical half of wings; the fork sometimes defined only weakly, by an apparent slightly oblique crossvein ▸ 3

3 Forewing costal crossveins not linked; vein CuP extending to about half length of the wing (fig. 147) ▸ ***Myiodactylus***
– Forewing costal crossveins linked in basal half of wing; vein CuP extending about one third length of the wing (fig. 148) ▸ ***Umbranymphes***

Notes on the genera

Osmylops Banks (type species *Myiodactylus placidus* Gerstaecker) is the most speciose genus in the family (Oswald, 1997, 1998a) with eleven described species. Three species occur in New Guinea, the others in Australia. The New Guinea species are all members of Oswald's (1998a) '*armatus*-group'.

Myiodactylus Brauer (type species *Myiodactylus osmyloides* Brauer) also occurs in both Australia and New Guinea, and the two faunas are very closely related. Two of the four described New Guinea species are known from single individuals from Papua (Irian Jaya). *M. nebulosus* McLachlan appears to be widely distributed in New Guinea, but more than one species may be masquerading under this name. *M. nebulosus* is separated at present from other species by lacking a continuous median stripe on the pronotum. The identity of the sole Philippines species is uncertain.

Umbranymphes New (type species *Umbranymphes spinosus* New) is known from a single species from Papua (Irian Jaya), which differs considerably in venational features from *Myiodactylus* but is otherwise probably closely related to it. On the form of the subcostal crossveins it can be confused only with *Norfolius* Navás, from eastern Australia and Lord Howe Island. In *Norfolius*, vein MP forks in the basal half of the wing.

Biological outline

It is presumed that the three genera mentioned above are arboreal, although there is little evidence for this except by analogy with Australian species. Larvae of *Osmylops* in Australia occur on tree foliage, and those of related species have been found on and under bark. Their highly flattened discoidal form as ambush predators is very characteristic. As with many lacewings, larvae appear to be polyphagous, but there are only few specific dietary records. Adults are usually captured at lights and are generally crepuscular or nocturnal.

MAJOR REFERENCES — New (1981): revision of Australian taxa; New (1983): early stages of *Osmylops*; Oswald (1997, 1998a): revision of *Osmylops*.

Ascalaphidae (owl flies)

149

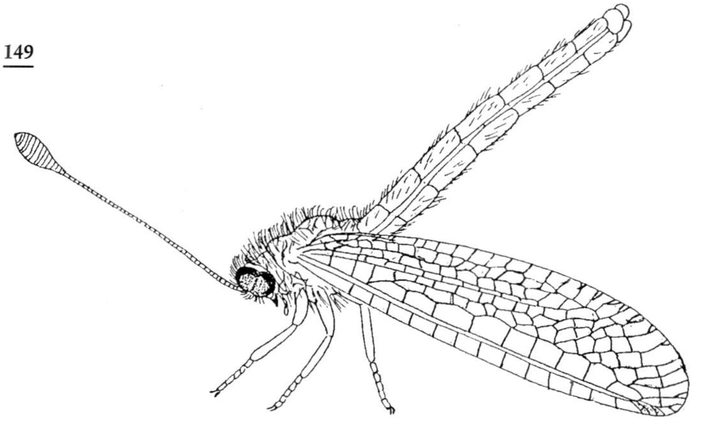

150

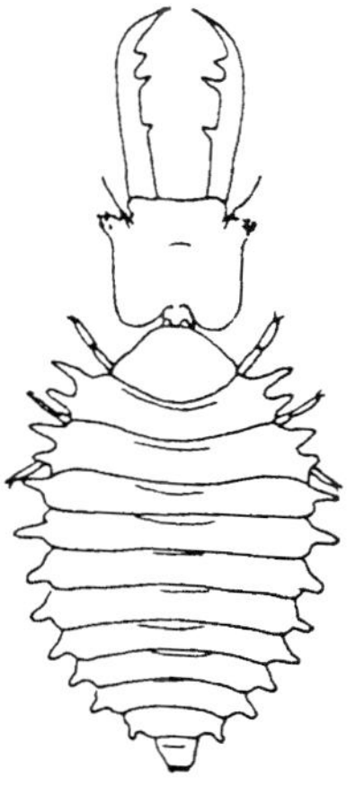

FIGURES 149 - 150
149 Adult Ascalaphidae
150 larval Ascalaphidae

Family diagnosis

Medium sized to large lacewings (forewing length approximately 15-60 mm), sturdy, often densely hairy. Head large, short, eyes very large; ocelli absent; antennae usually very long, flagellum slender except for well defined club at apex, flagellum rarely sinuous or ornamented near base in males; mandibles pointed; palpi usually short. Pronotum reduced, transverse. Pterothorax enlarged. Legs short and sturdy, often strongly spined; tibial spurs present; tarsal claws strong, simple. Wings (fig. 151) elongate, subequal or with hindwing shorter and broader than forewing; trichosors absent; pterostigma usually well defined, often small and dark; membrane hyaline or variously marked. Costal space narrow; numerous crossveins, usually simple, no recurrent humeral crossvein; Sc and R1 fused apically, generally no Sc crossveins; Rs and MA with common stem; Rs pectinate. Forewing major fork on posterior margin from CuA; hindwing major fork on posterior margin from MP2; forewing CuP and 1A distinct, these veins fused in hindwing; anal angle of forewing sometimes produced or angled. Abdomen elongate, usually sturdy; male sometimes with prominent dorsal projection (simple or furcate) on tergite i, ii, or iii; female abdomen not ornamented.

FIGURES 151 - 159

Ascalaphidae - **151** general pattern of wing venation (major longitudinal veins, only, shown; a, apical field) — general pattern of abdominal apex and genitalic structures of **152** female in lateral aspect, **153** of female in ventral aspect, **154** of male in lateral aspect, **155** of male in lateral aspect (dv = dorsal gonapophysis valve; ect = ectoproct; g = gonosetae; go = gonarcus; li = linguella; pa = parameres; pe = pelta; pv = pulvinus; vv = ventral gonapophysis valve) (from New, 1984) — **156** eyes entire, as in Haplogleniinae, or **157** divided by horizontal sulcus as in Ascalaphinae — **158** male ectoproct elongate (*Ascalohybris*) or **159** rounded (*Suhpalacsa*).

151

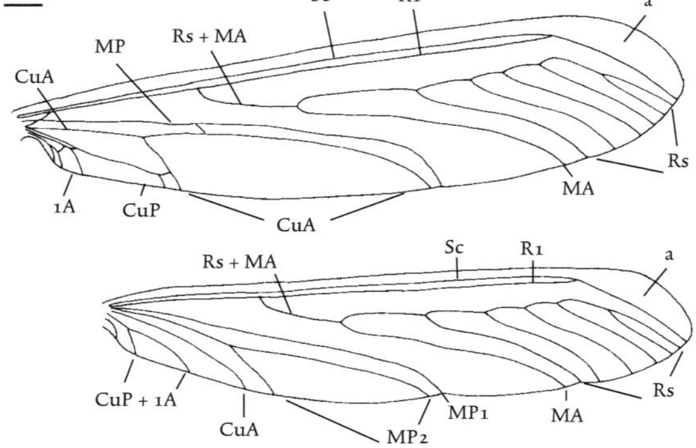

152

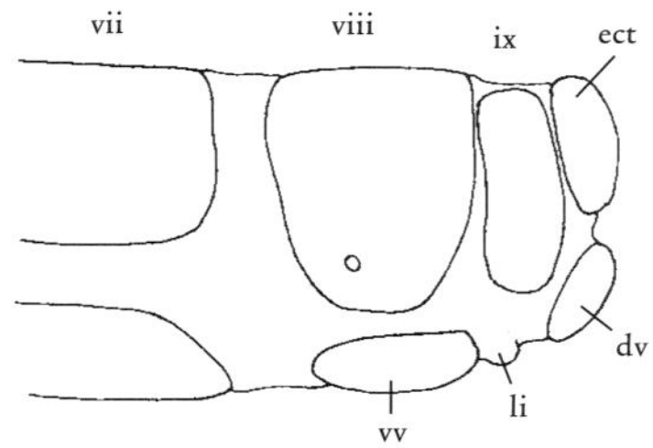

153

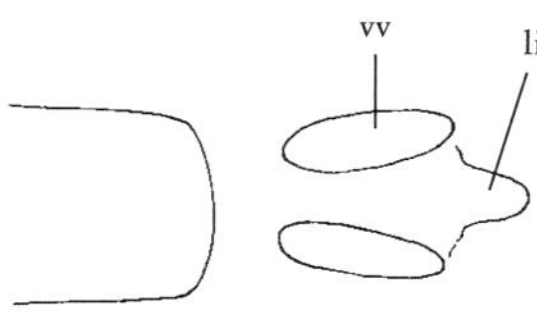

154

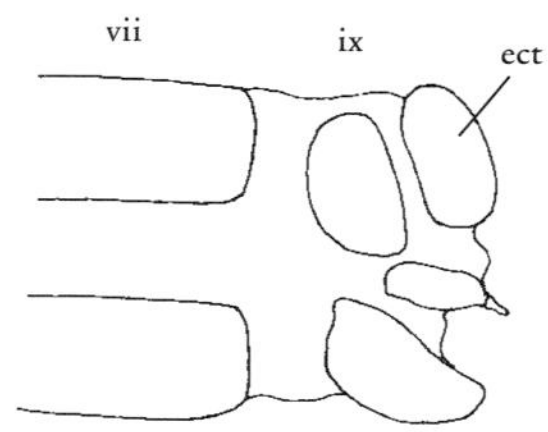

155

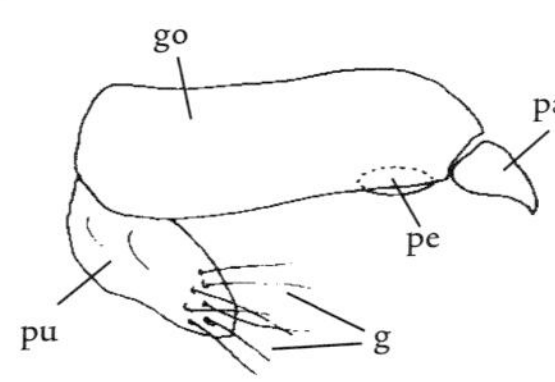

156

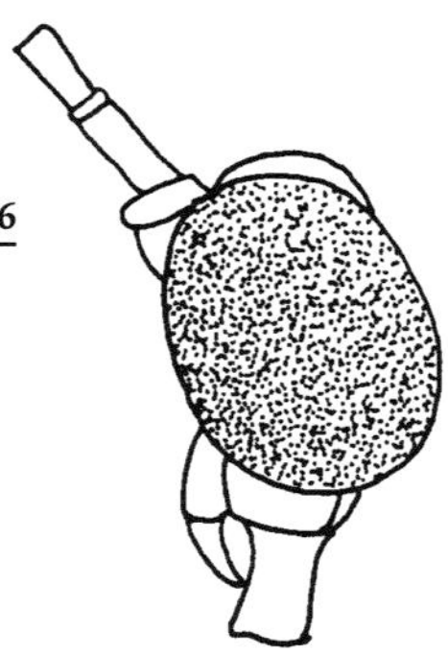

157

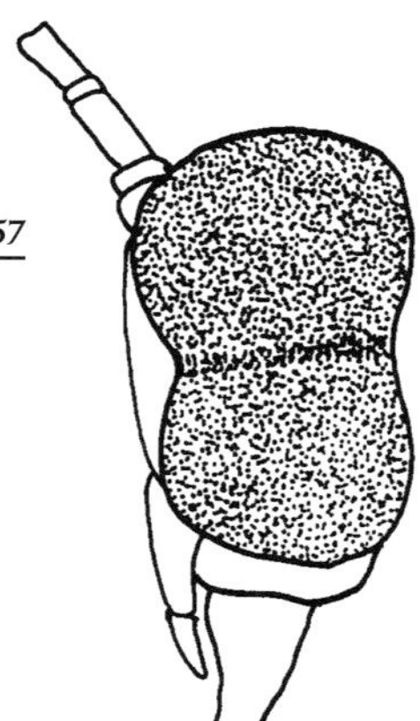

158

159

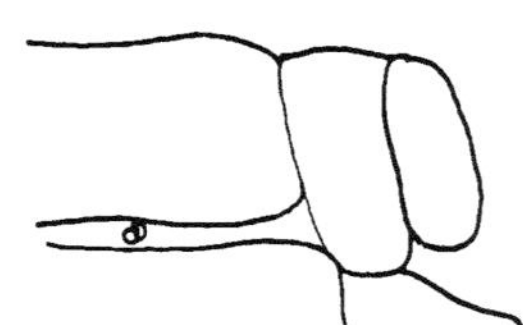

Female (figs. 152, 153): tergite viii large; tergite ix usually slender; sternite viii divided to form ventral gonapophysis valves and distal median linguella; lateral gonapophyses short, unornamented; spermatheca small, tubular.
Male (figs. 154, 155): tergite ix small; sternite ix usually simple; ectoprocts simple or forcepate; gonarcus elongate, apex with small hooked parameres and ventral pelta, and lateral setose membranous lobes (pulvini); hypandrium internum small.

Diversity and distribution

There are about 380 species of Ascalaphidae (25 in Malesia), commonly termed 'owlflies' because of their very large eyes. They are very distinctive Neuroptera, sometimes likened to dragonflies with long antennae, usually, with a very distinctive apical club. They are fast flying vigorous predators. They are generally considered to be related to Myrmeleontidae, and the larvae of the two families are sometimes difficult to differentiate. Adults of the two families differ in many features, but perhaps most obviously in the long antennae of most ascalaphids. The only Ascalaphidae with relatively short antennae are the rare South American Albardiinae, which were formerly placed in the composite family Stilbopterygidae together with some unrelated Australian myrmeleontids (New, 1982). Some Myrmeleontidae have antennae with a well-defined club, but in such species the antennae are always short and do not exceed about one-fifth of the length of the forewing. The venation also differs substantially between the two families.
The two main subfamilies of Ascalaphidae are both widespread. As Tjeder (1992) commented it is remarkable that the only fundamental character found to support division into these subfamilies is that of the compound eyes rather than any clearcut differences of wing venation or genitalia, both of which have considerable values in species and generic separations. However, Tjeder (1992) noted that a pleurostoma (a small triangular plate between the gena and the base of the mandible) was present in all Ascalaphinae he examined and absent in all Haplogleniinae except *Proctolyra* Tjeder, a southern African genus possibly annectent between the subfamilies. Ascalaphinae are considered the more advanced subfamily.

Key to main subfamilies of Ascalaphidae

1 Eyes 'entire', that is not divided into two regions by a transverse furrow or sulcus (fig. 156) ▸ **Haplogleniinae**

– Eyes 'split', that is divided superficially by a transverse furrow or sulcus (fig. 157) ▸ **Ascalaphinae**

Notes on the subfamilies

Haplogleniinae is by far the less diverse subfamily in Malesia and is absent from Australia, whilst Ascalaphinae is widespread in the region and includes the entire Australian ascalaphid fauna of 36 described species (New, 1984). The two groups were considered as separate families by Newman (1853), but this opinion has never been adopted widely and is now universally discounted. Both subfamilies have been known by a variety of other names during their history, some of these being invalid as not being based on any described genus. Thus, the names 'Holophthalminae' and 'Schizophthalminae', used by Van der Weele (1909b), refer to the above conditions of the eyes, but are invalid names. Haplogleniinae have four major centres of distribution, with the generic ranges each wholly confined to a single range centre. This renders regional assessment at one level straightforward, but has deterred global evaluation of the taxa in favour of regional studies so that the relationships (and possible synonymies) between rather similar genera in different parts of the world remain unclear. The range of interest here is southern Asia (from the Near East through India to southern China), including also Japan, Borneo and the Philippines, as one natural range for a group of genera. Although Navás (1913) described a species of the American genus *Haploglenius* from New Guinea, this record has been doubted by later workers. To quote from Tjeder (1992): 'from a zoogeographical point of view the record seems hardly credible'. Tjeder suggested the possibility that the specimen had been labelled wrongly. Major groupings in the subfamily (tribes) were noted by Van der Weele (1909b) in his global monograph of the family (which, although dated, is still the most fundamental work on the family), and that arrangement has scarcely been changed for Malesian taxa. All Malesian representatives have distinctly clubbed antennae, serving to differentiate them from the African Tmesibasini in which the apical antennal segments are scarcely thickened. The few species are at present allotted to two genera, differentiated on the shape of the forewing base.

Ascalaphinae are also varied, but the great majority of species from Southeast Asia belong to the tribe Ascalaphini (including the Suhpalacsini of Van der Weele 1909, and many later authors including New (1984)), characterised by the male having short scale-like ectoprocts (fig. 159). The ectoprocts of males of other tribes in the region and Australia are forcepate (Encyoposini), or enlarged but not markedly forcepate (Acmonotini, although this tribe is sometimes included in Ascalaphini, following Navás 1919). Earlier tribal allocations used by Van der Weele (1909b) are included in the key to facilitate comparison with earlier literature.

In both subfamilies, generic limits are not always well defined, and the Ascalaphini are particularly complex, with most genera very poorly delimited. Only since Tjeder & Hansson's (1992) appraisal of the Afrotropical taxa, for example, has it become clear that all those regional taxa are generically distinct from Asian and Australasian forms, because some African species were previously placed in *Suphalomitus* or other extralimital genera. In addition, as appraisal of the Australian ascalaphids (New, 1984) demonstrated, the great variety of forms in the tribe has led to proliferation of small genera on small (but apparently valid) characters. As many of the Malesian species have not been examined critically, their precise generic position according to modern concepts is simply unknown, and the genera listed are those to which species have been attributed in the literature, predominantly in original descriptions. In particular, some species placed in *Suhpalacsa* Lefèbvre or *Suphalomitus* van der Weele may need to be transferred elsewhere, and the following key characters are thus deliberately broad to incorporate wider 'historical' concepts of these genera. One widespread species, treated most recently as *'Ascalaphus' dicax* Walker by Stange & Wang (1998) is more commonly referred to in regional literature as *Helicomitus* McLachlan, considered as a synonym of *Ascalaphus* Fabricius by Tjeder (1972).

FIGURES 160 - 170

Ascalaphidae: wing venation of **160** *Protidricerus*, **161** *Idricerus*, **162** *Glyptobasis*, **163** *Helicomitus*, **164** *Protacheron*, **165** *Suhpalacsa*, **166** *Suphalomitus*, **167** *Ascalohybris*, **168** *Nousera*, forewing base, **169** *Nousera*, hindwing base (figs. 168-169 after Navás (1923), original description) — **170** basal antennal flagellar segments of *Acheron* with lateral spines.

160

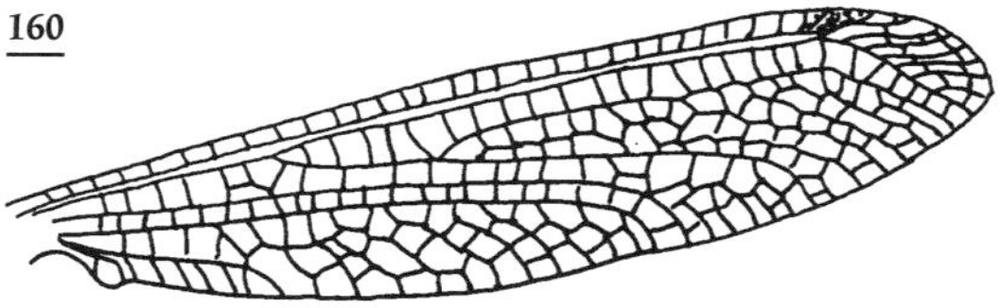

161

162

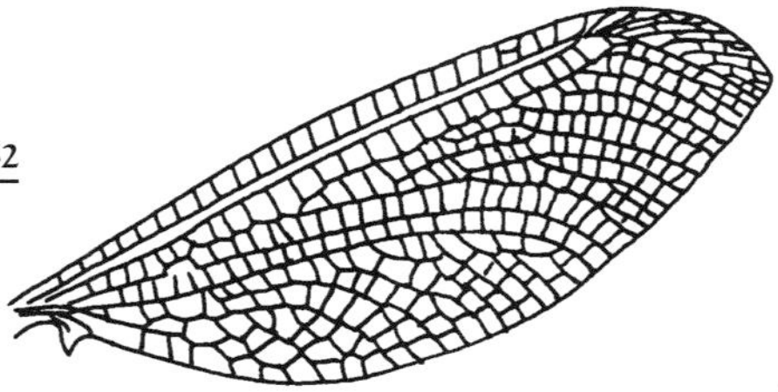

163

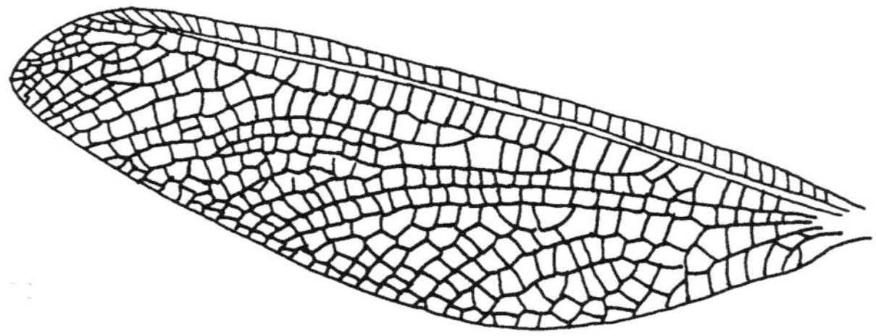

164

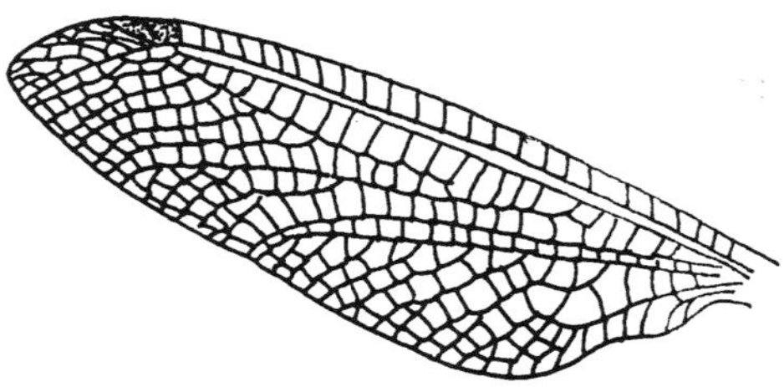

165

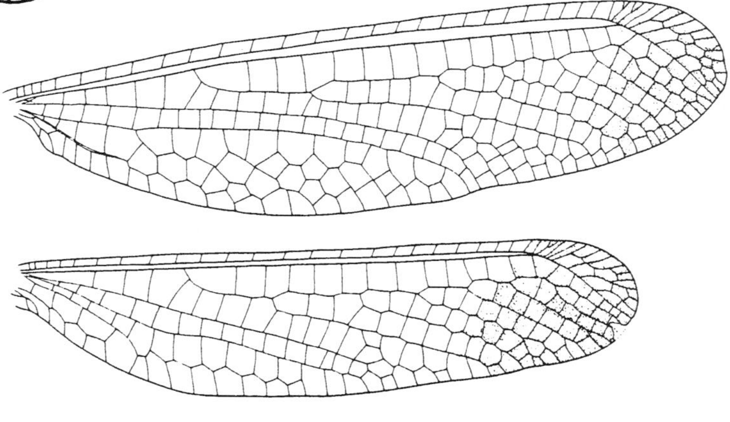

166

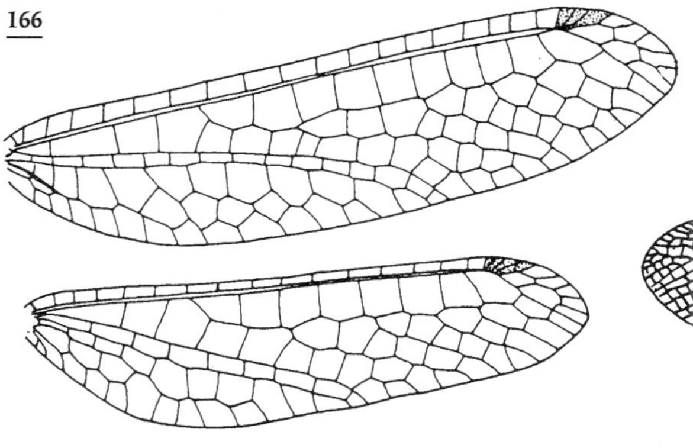

167

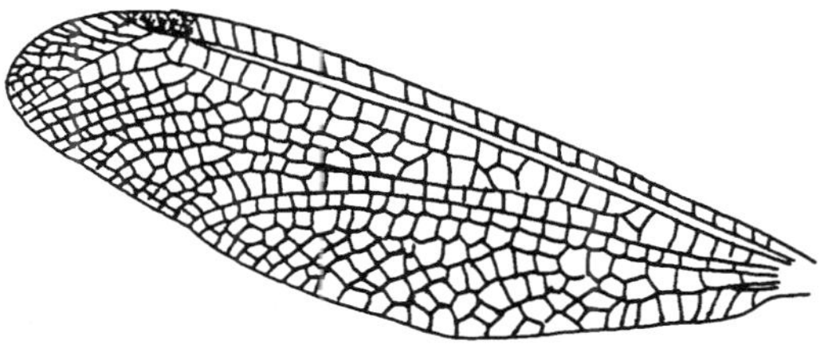

168

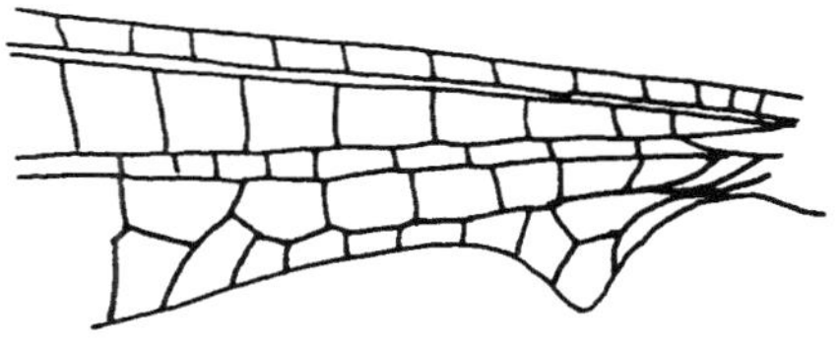

169

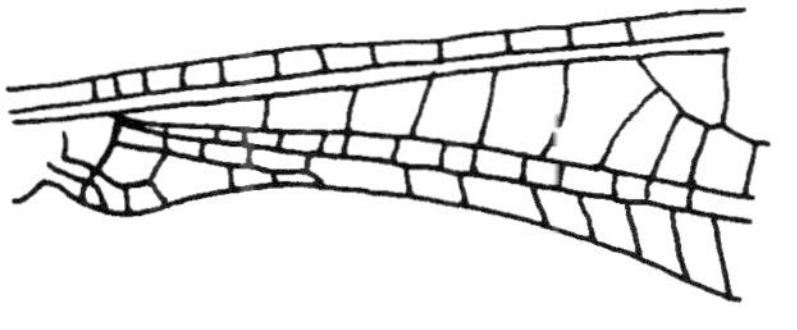

170

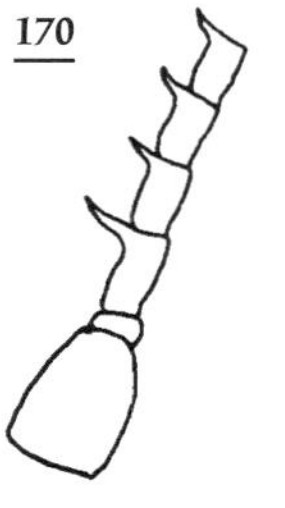

Key to the genera of Haplogleniinae

The two regional genera were separated by Van der Weele (1909b) predominantly on the shape of the forewing base.

1 Forewing base with rounded or semicircular posterior excavation (fig. 160); adjacent hind margin scarcely excavated; anterior and posterior margins largely parallel ▸ *Protidricerus*

– Forewing base with rounded or semicircular posterior excavation, followed by a broad, blunt posterior projection and deep invagination on the posterior margin (fig. 161) ▸ *Idricerus*

Notes on the genera

Idricerus McLachlan (type species *Ascalaphus decrepitus* Walker) is included here on the basis of one record (*I. obscurus* (Westwood)) from 'East Indies'; the other species of the genus occur in China or India, and confirmation of this record is needed.

Protidricerus van der Weele (type species *Idricerus exilis* McLachlan). Species of *Protidricerus* have been reported from Borneo and the Philippines, as well as India, China and Japan. One species is 'fairly common' in Taiwan (Stange & Wang, 1998), but species level taxonomy is in need of revision. The forewing pterostigma is very short and the hindwing pterostigma generally somewhat longer.

Key to the genera of Ascalaphinae

The nominal genera reported from Malesia and nearby regions are separated tentatively as follows. Tribal divisions are also subject to revision.

1 Basal antennal segments of male with distinct lateral spines; large species: forewing length c. 45 mm ▸ *Acheron*
– Basal antennal segments of male without lateral spines; species generally smaller, forewing length usually less than 40 mm ▸ 2

2 Abdomen of male (or of both sexes) with ectoprocts extended ventrally, sometimes forcepate (fig. 158) ▸ (Hybrisini of Van der Weele) 3
– Abdomen with ectoprocts short, rounded (fig. 159) ▸ (Suhpalacsini of Van der Weele) 4

3 Male ectoprocts relatively short, not protruding markedly ▸ *Protacheron*
– Male ectoprocts with long protruding process, flexed ventrally ▸ *Ascalohybris*

4 Forewing base with extended triangular axillary lobe ▸ 5
– Forewing base without pronounced triangular axillary lobe, rounded or slightly expanded ▸ 6

5 Pterostigma deeper than long ▸ *Nousera* (not in Malesia?)
– Pterostigma longer than deep (fig. 162) ▸ *Glyptobasis*

6 Both wings broadened considerably at about 1/3-1/2 length, then narrowed toward apex (fig. 163, see also fig. 164) ▸ ***Helicomitus***

– Both wings narrow, straplike, anterior and posterior margins almost parallel ▸ 7

7 Pterostigma short, about as long as deep; hindwing of similar width to forewing, usually little shorter than forewing (fig. 165) ▸ ***Suhpalacsa***

– Pterostigma long, about twice as long as deep; hindwing broader than forewing, relatively short and usually not extending beyond forewing pterostigma (fig. 166) ▸ ***Suphalomitus***

Notes on the genera of Ascalaphinae

Helicomitus McLachlan (type species *Ascalaphus insimulans* Walker) is, as noted above, often considered a synonym of *Ascalaphus*, but further investigation of this is needed. The sole species in Malesia, '*H.*' *dicax* (Walker), appears to be widespread. The possible distinction of *H. placidus* (Gerstaecker) and *H. dicax* needs confirmation; they were listed separately by Sziraki (1999).

Ascalohybris Sziraki, a replacement name necessitated by homonymy of *Hybris* Lefèbvre (type species *Ascalaphus javanus* Burmeister) may be the most speciose genus of ascalaphids in Malesia, though most of its 14 nominal species have been characterised poorly and some are only doubtfully distinct. Existing records suggest that *Ascalohybris* is most diverse in Indonesia (with five species), within a broader generic range from India to China and Taiwan. The greatly extended male ectoprocts are diagnostic in the regional fauna. Many of the regional species have a broad mid-dorsal yellow stripe along the thorax and abdomen (wing fig. 167).

Nousera Navás (type species *Nousera gibba* Navás; figs. 168, 169) was raised for a single species from Indochina, which was not reappraised since it was described by Navás (1923) until recently (Sziraki, 1999). Navás likened *Nousera* to *Acmonotus* McLachlan, but it differs by lacking a dorsal abdominal projection in the male and in the more pronounced forewing axillary lobe. The latter feature should afford easy recognition, should the genus be found in Malesia. Sziraki (1999) noted the synonymy of *Nousera* with *Pseudoptynx* van der Weele. Because of homonymy of the latter it was replaced formally by its new junior synonym. The axillary lobe of *Glyptobasis* McLachlan (fig. 162) is also well developed, but the pterostigma is longer than deep; this genus occurs in Taiwan and the Indian subcontinent, but has not been reported from Malesia. However, two species occur in Myanmar, and it may extend into the northern parts of Malesia.

Acheron Lefèbvre (type species *Ascalaphus longus* Walker). *Acheron trux* (Walker), the sole widely recognised species of the genus, is regarded as the most widespread ascalaphid in Taiwan (Stange & Wang, 1998). Within Malesia, it is known from West Malaysia. It is known also from China and India, and should be sought in other more northerly parts of Malesia, such as the Philippines. *Acheron* is recognisable by the basal antennal flagellar segments of the male having lateral spines (fig. 170). These are often reduced in females, but *A. trux* females (with forewing length around 45 mm) are among the largest ascalaphids in the region. Malaysian specimens sometimes have the costal area of both wings dark brown, while many specimens from elsewhere have the wings wholly pale. More than one species may be present in this genus, and several supposed synonyms need objective confirmation of this status. Ghosh (1988) recognised two subspecies of *A. trux*: *A. t. trux* and *A. t. loquax* (Walker), originally described as a separate species but placed in synonymy by later workers.

Protacheron van der Weele (type species *Helicomitus philippinensis* van der Weele) is known by two species. The type species has been reported from Sulawesi and Java as well as the Philippines. It is moderate-sized (forewing length approximately 28 mm) and has unmarked wings. *P. westermanni* Esben-Petersen occurs in the Himalayas.

Suhpalacsa Lefèbvre (type species *Ascalaphus flavipes* Leach) is most diverse in Australia, but occurs also in Papua New Guinea, Indonesia and the Philippines, with the genus extending as far west as India. Males of some Australian species have a dorsal prominence on the anterior part of the abdomen, but this feature has not been noted in Malesian species. Many of the species are very similar in general appearance. Most have unmarked wings, but *S. dietrichiae* (Brauer) usually has the wing apices suffused with pale brown. *S. princeps* (Gerstaecker) and another large species from Malaysia have parts of the costal area dark brown. The names '*Suhpalasca*', '*Suphalacsa*' and '*Suphalasca*' all occur in the literature as incorrect spellings.

Suphalomitus Van der Weele (type species *Suhpalacsa difformis* McLachlan) is closely related to *Suhpalacsa*. It differs consistently on the relative length of the wings and the shape of the pterostigma.

Biological outline

Some owlflies are diurnal, whilst others are crepuscular and come to light. Diurnal species may have relatively broad and brightly patterned wings, in contrast to nocturnal species, in which wings are usually narrow and unmarked. They are among the most vigorous and strongly flying Neuroptera. Some species show pronounced sexual dimorphism, with the males having more elaborate antennae than females, or (some Ascalaphinae) well developed dorsal projections on the anterior part of the abdomen. Larvae intergrade in appearance with those of Myrmeleontidae, but usually have a more strongly excavated posterior head margin and more pronounced lateral body processes. They are found in ground litter and on or under bark of trees, and may become covered with debris. Eggs are laid in batches (commonly, of 30-60 eggs) on twigs or other vegetation, and repagula appear to be absent from regional taxa.

MAJOR REFERENCES — Van der Weele (1909b): world revision and synopsis of family; Henry (1972): form and functions of repagula; Henry (1976): larval morphology; New (1984): revision of Australian fauna; Tjeder (1992): revision of African forms, with much general background on family; Sziraki (1999): list of Asian and Pacific species, with some new generic replacement names.

Myrmeleontidae (antlions)

171

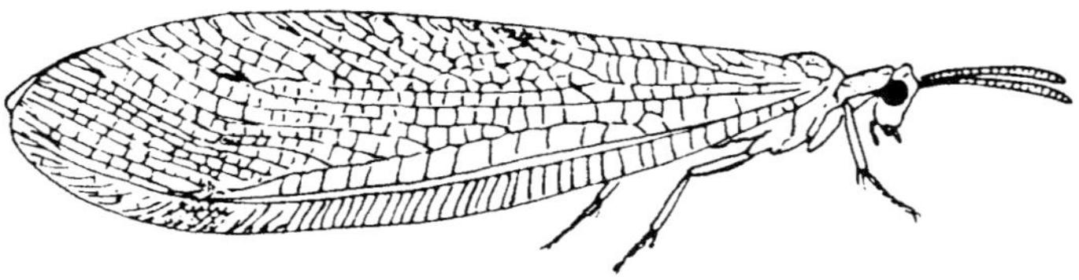

172

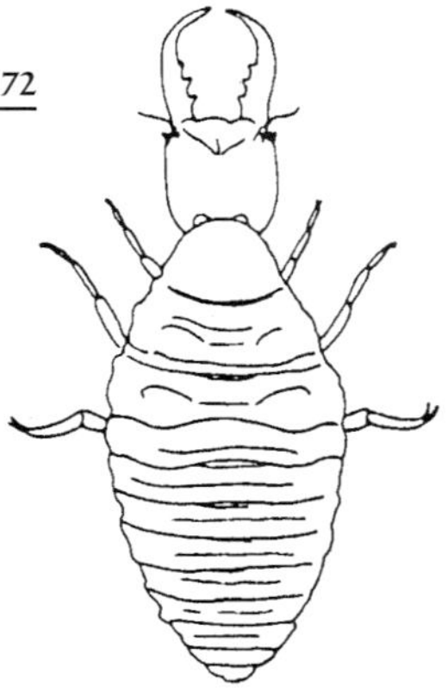

FIGURES 171 - 172
171 Adult Myrmeleontidae
172 larval Myrmeleontidae

Family diagnosis

Medium sized to very large lacewings (forewing length approximately 10 to over 70 mm), slender to very sturdy, sometimes densely hairy. Head moderately large; eyes large; ocelli absent; antennae short, always less than half the length of the forewings, usually thickened or incipiently clubbed near apex. Mandibles symmetrical. Pronotum well sclerotised, often with transverse furrow; mesonotum often longer than metanotum. Legs slender to sturdy; tarsal claws simple or toothed; femora and tibiae often spined, tibial spurs present or absent. Wings subequal, usually much longer than wide, apex tapered or rounded, rarely falcate; trichosors absent; pterostigma well defined; membrane hyaline or variously patterned. Costal area narrow, with many crossveins, humeral crossvein simple; Sc and R1 fused distally, usually lacking basal Sc-R1 crossveins; Rs and MA arise on common stem, usually few presectoral crossveins; Rs with several branches, number of gradate veins highly variable, branches sometimes bent to form almost straight line (anterior Banksian line; fig. 1) along centre of wing; forewing major fork in posterior half formed by CuA, in hindwing major fork formed by MP2; posterior Banksian line sometimes formed by bending of crossveins within these forks; linking of 2A and 3A by crossveins or central length fusion; hindwing of male sometimes with basal posterior knobbed lobe associated with glandular pocket on first abdominal segment. Abdomen elongate, usually slender, sometimes sturdy.
Female (figs. 173, 174): ectoproct simple; tergites viii and ix ventrally expanded; ectoprocts and rounded lateral gonapophyses sometimes with conspicuously thickened 'digging setae'; posterior gonapophyses at inner basal margin of lateral gonapophyses; spermatheca slender.

FIGURES 173 - 176
Myrmeleontidae - schematic arrangement of abdominal apex and genitalic structures to indicate terminology: female, apex of abdomen **173** in lateral aspect and **174** in ventral aspect; male, **175** apex of abdomen, lateral aspect, **176** genitalic complex, lateral aspect (segments numbered vii-ix; ect = ectoproct; a.g. = anterior gonapophysis; l.g. = lateral gonapophysis; p.g. = posterior gonapophysis; p.p. = pregenital plate; S. = spermatheca; gen. = male genitalic complex; gon = gonarcus; med = mediuncus; par = parameres) (from New 1985a).

173

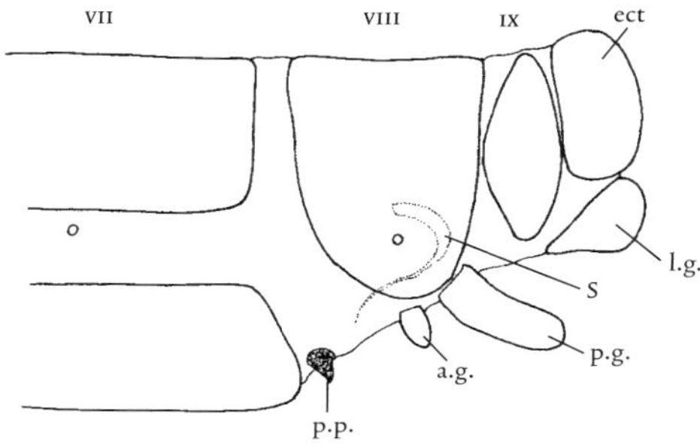

174

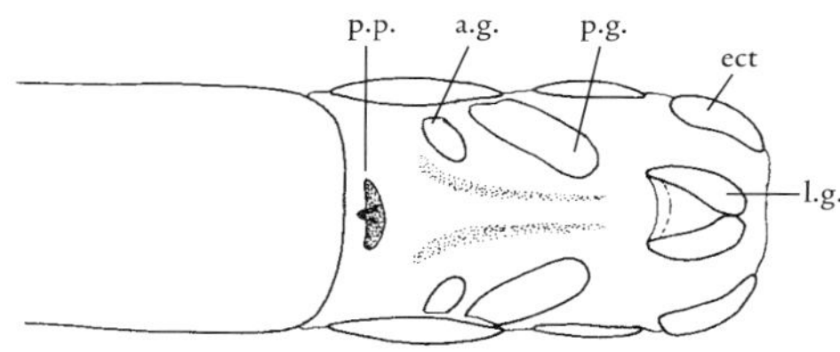

175

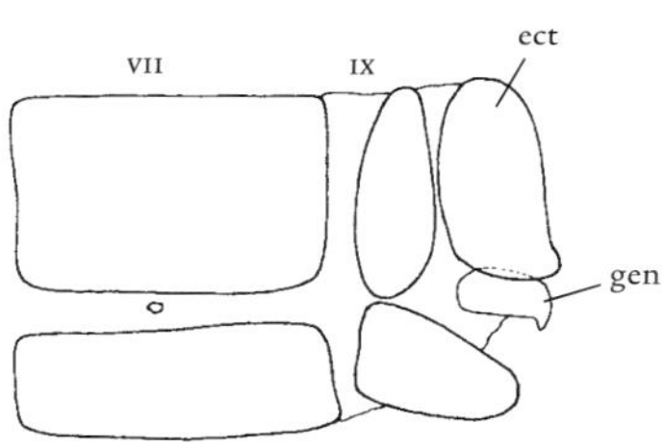

176

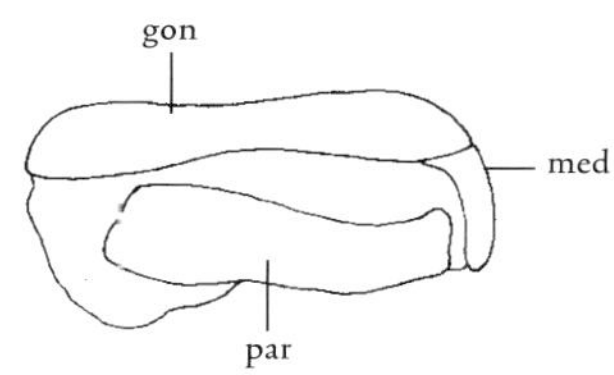

Male (figs. 175, 176): tergite ix closely associated with ectoproct; ectoproct sometimes ornamented and/or extended ventrally; sternite ix small, often rounded; gonarcus arched; mediuncus present; pair of short parameres present; hypandrium internum small, usually indistinct.

The higher taxonomy of Myrmeleontidae is complex and still somewhat unsettled (Mansell, 1999), so that a considerable variety of subfamily and tribal categories have been erected since Banks (1899) distinguished two groups, the Myrmeleoni and Dendroleoni, later elevated (Banks, 1911) to subfamilies as Mymeleoninae (with tribes Myrmeleonini and Palparini) and Dendroleoninae (with tribes Nemoleonini and Dendroleonini). A somewhat different primary division by Esben-Petersen (1918) separated Archaemyrmeleonida (containing only Palparini) from Neomyrmeleonida (Myrmeleontinae [with four tribes] and Dendroleontinae [with six tribes]), and several of those tribes were treated by later authors as subfamilies.

The most complex and embracing higher classification, by Markl (1954), separated the genera into 23 tribes, with no consideration of possible subfamily groupings. As Mansell (1999) commented, this was a highly significant basis for future discussions. A variety of other arrangements has ensued, but none of the 'modern subfamilies' is based on sound phylogenetic study (Mansell, 1999), so debate on the optimal higher classification in the family is likely to continue. The major groups of relevance to the Malesian region are Acanthaclisinae and Myrmeleontinae (with Myrmeleontini, Dendroleontini, Protoplectrini, Distoleontini), though Mansell (1999) provided support for a two-clade concept in the family, so that the above would be modified to subfamilies Myrmeleontinae and Dendroleontinae (cf. Banks, 1899!), with Acanthaclisini being a tribe of Myrmeleontinae. For users of this account, it is simply relevant to note that higher classification in Myrmeleontidae has undergone considerable changes over the last century, that broad consensus is still to be reached, and that a variety of higher group names have been applied in the literature to the taxa discussed here.

Many of the species have not been appraised critically, so that some generic allocations are very tentative as having been made many years ago when ideas on generic limits were not as well-defined. Indeed, despite their large size and conspicuous appearance, the antlions are the most confusing group of lacewings in the region, and the affinities of many taxa, particularly some described by Navás, need clarification.

The spectacular Palparini (or Palparinae: Hölzel 1972) are virtually absent from the region, if the isolated species of *Palpares* Rambur and *Stenares* Hagen described from 'East Indies' are accredited with a misinterpretation for 'East India', within the extended range of these predominantly African genera. The two genera are closely related (Mansell, 1992). *Palpares pardus* Rambur occurs in India and Myanmar, and true *P. spectrum* Rambur is a member of a species complex, all of them African. The pertinent Walker specimen of *S. improbus* (Walker) appears to have been lost. The Indian *S. improbus* is a large species (forewing length approximately 60 mm) with very dense venation and considerable brown spotting, especially toward the posterior margin of the hindwing. Neither genus has been found in Malesia since the species were described. Another anomaly is Van der Weele's (1909a) record of a specimen of the northern Indian *Tomatares pardalis* (Fabricius) from Java. This specimen (labelled 'Java, Horsfield') is in the Natural History Museum, London, and does not differ in any essential superficial details from Indian specimens. I am not aware of any more recent records of this striking insect in the region, but it would be recognisable readily by the short, strongly clubbed antennae and broad yellow wings with numerous black or dark brown transverse lines (forewing) and spots, including a large central spot (hindwing). *Tomatares* contains four species (Mansell, 1992) and is predominantly African. The Palparinae is probably represented in Malesia solely by *Echthromyrmex*, although the above records can not be discounted without further investigation.

Diversity and distribution

The Myrmeleontidae are the largest family of Neuroptera. About 2000 species are currently recognised, of which 60 have been recorded from Malesia, and the family includes the largest and most visually striking lacewings. Wing spans range up to around 170 mm, with wings of some species ornately and intricately patterned. As Mansell (1999) noted, Myrmeleontidae are a highly evolved family, whose evolutionary diversity has been fostered by a great variety of predatory strategies developed in the larvae. The name 'Myrmeleontidae' has now been adopted generally as etymologically correct and is therefore preferable to the widely used alternative 'Myrmeleonidae' (Canard et al., 1992).

The richest radiations of Myrmeleontidae tend to be found in semi-arid regions, such as much of Australia, inland southern Africa and southern North America. The Malesian region seemingly supports rather few species. Myrmeleontinae are very widespread and occur on all continents. Dendroleontinae are also widely distributed, but have undergone greatest radiation in Australia. The Malesian taxa have close affinities with Australasian forms. Acanthaclisinae (or Acanthaclisini) are distinctive and also represented on all continents, but by different genera on each, so that considerable regional endemism at this level occurs.

Key to higher taxa of Myrmeleontidae in Malesia

Terminology for wing venation follows New (1985a, b, c), and major features are summarised in figs. 177-184; the basic pattern yields many features of taxonomic value, as reflected in this key. Wing shape may also have generic value in some groups, and characters of legs, claws and spurs, and genitalia of both sexes are used widely in species diagnoses and generic recognition.

1 Hindwings with three complete anal veins (fig. 178) ‣ **Palparinae** (p. 129)
– Hindwings with two anal veins ‣ 2

2 Femoral sense hair often present on hindlegs; generally large, densely hairy species with more than four presectoral crossveins in hindwing; forewing costal cells partially or completely biareolate (figs. 185, 186). (Legs short and stout; tarsal claws and tibial spurs very strongly arched.) ‣ **Acanthaclisinae** (p. 129)
– Femoral sense hair not present on hindlegs, sometimes present on anterior legs; body usually not densely hairy; 1-6 presectoral crossveins in hindwings; if more than two, forewing costal cells simple except (rarely) immediately basal to pterostigma (fig. 192). (Legs usually slender, tarsal claws and tibial spurs of various forms.) ‣ **Myrmeleontinae** 3

3 Veins 1A and 2A clearly separable at base of forewings; veins 2A and 3A linked by crossvein, meeting at a point, or adjacent for a short distance (fig. 181) ‣ **Dendroleontini** (p. 138)
– Veins 1A and 2A very close together at base of forewings; veins 2A and 3A closely associated over central length (fig. 182) ‣ 4

4 Hindwing vein Rs arising opposite or beyond medial fork; usually four or more presectoral crossveins in hindwings (fig. 183) ‣ **Myrmeleontini** (p. 132)
– Hindwing vein Rs arising well before medial fork; one, rarely two or more, presectoral crossvein(s) in hindwings (fig. 184, 185) ‣ 5

5 Forewing veins CuA2 and (CuP + 1A) elongate, running parallel to CuA1 and hind margin of wing for a considerable length (fig. 197) ‣ **Protoplectrini** (p. 134)
– Forewing veins CuA2 and (CuP + 1A) not as above, short (fig. 205) ‣ **Distoleontini** (p. 134)

FIGURES 177 - 184

Myrmeleontidae - **177** general pattern of wing venation to indicate terminology (after New, 1985a) — generalised Palparini, basal region of **178** hindwing and **179** forewing, (note 1A meets CuP branch in forewing) — **180** hindwing, *Echthromyrmex*, **181** forewing of Dendroleontini (2A and 3A linked by crossvein), **182** forewing of Myrmeleontini (2A and 3A closely associated in central region), hindwing of **183** Dendroleontini or Distoleontini with single presectoral crossvein, **184** Mymeleontini with several presectoral crossveins.

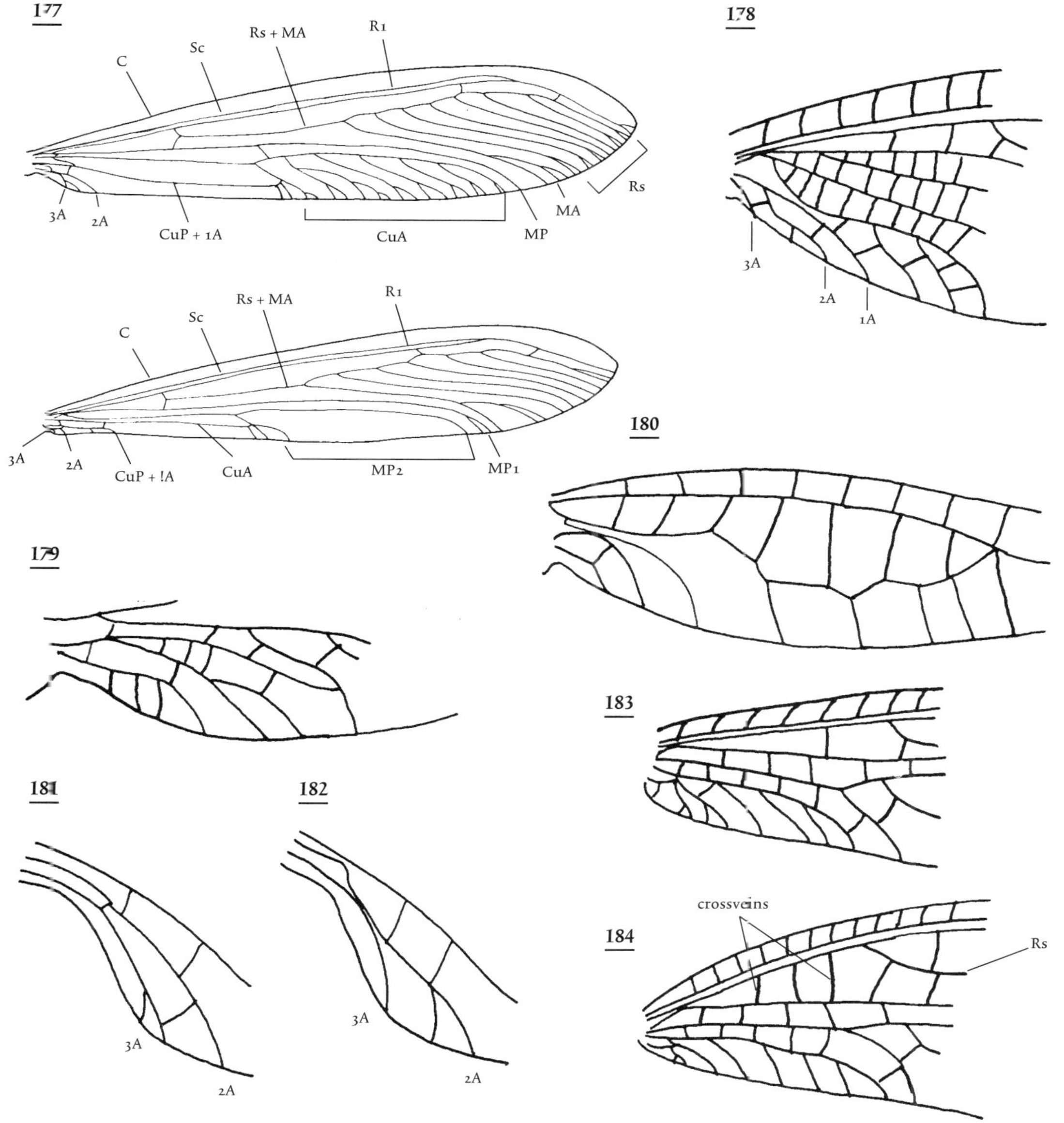

FIGURES 185 - 190

Myrmeleontidae - Acanthaclisinae: wing venation of **185** *Cosina*, **186** *Heoclisis* — **187** tibial spurs strongly angled or **188** shallowly curved — **189** general form of abdominal apex of male and **190** of female.

185

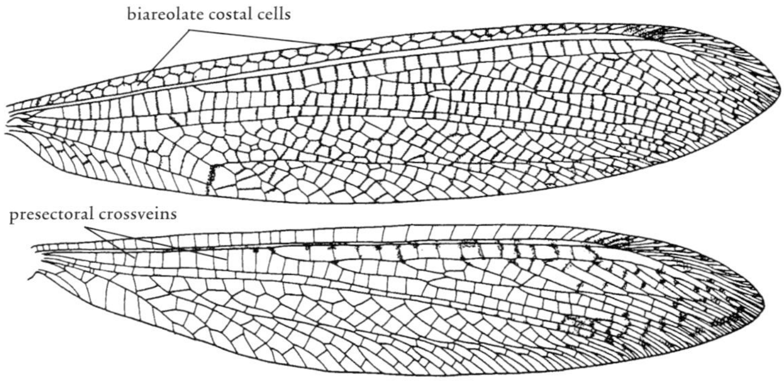

186

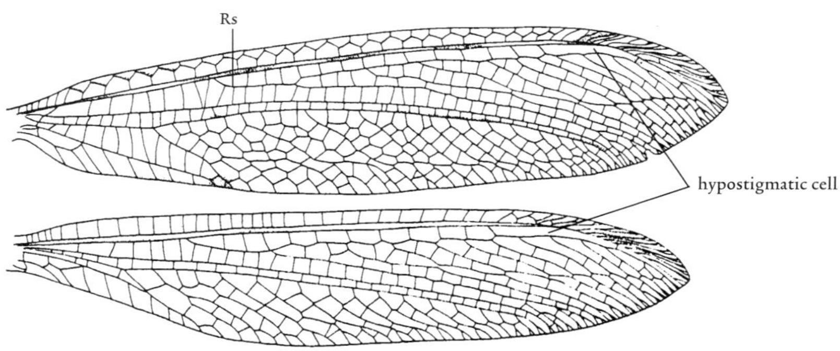

187

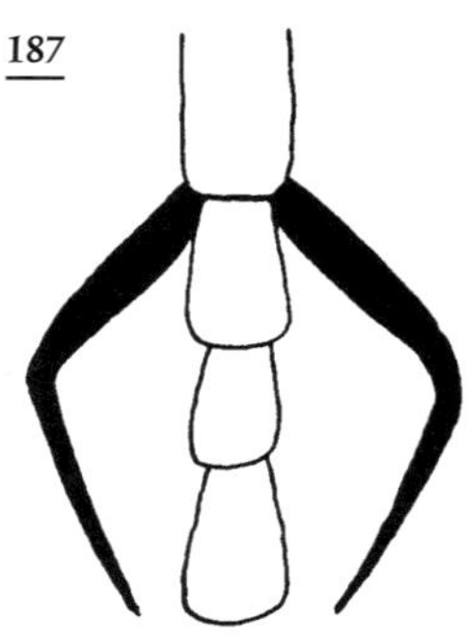

188

190

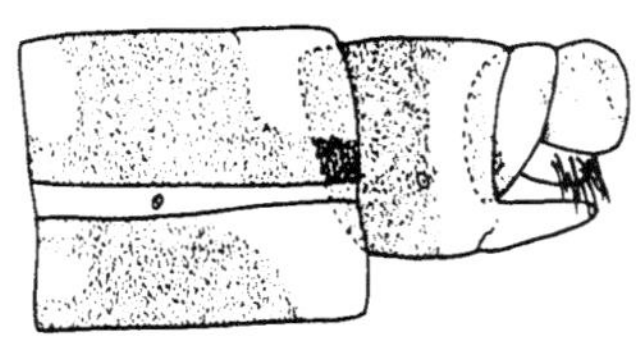

189

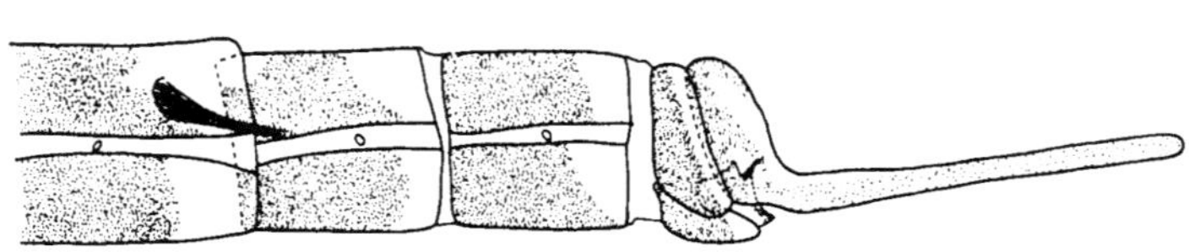

Notes on the genus of Palparinae

Echthromyrmex McLachlan (type species *Echthromyrmex platypterus* McLachlan) is now generally considered to represent a tribe (Echthromymicini by itself or, more probably, one of three genera in Dimarini; Stange, 1989) of Palparinae, rather than forming an independent subfamily (Echthromyrmicinae) as suggested by Hölzel (1972). It differs from all other Palparinae in having forewing vein CuP partially fused with 1A (fig. 180), rather than running separately to the wing margin (Stange & Miller, 1990). The single species in the region, of only four described, is *E. orientalis* McLachlan, known from Myanmar and the Moluccas (see Hölzel, 1986). It is a large antlion (forewing length 45-50 mm) with both wings having strong brown bands across the apical and preapical region. The other species are found in Iraq, Socotra and South Africa. See notes preceding the key on higher groups of Myrmeleontidae on other possible Palparinae that could be found in the region.

Key to genera of Acanthaclisinae

NOTE
Wing shape differences quoted after Stange & Miller (1985): some Australian species referred to *Heoclisis* have slight irregularities in wing shape, and this feature may not be wholly reliable.

1	Tibial spurs bent at nearly a right angle, with a 'flange' before the bend (fig. 187) ▸	*Acanthaclisis* (not in Malesia?)
–	Tibial spurs shallowly curved (fig. 188) ▸	2
2	Hindwings with vein Rs arising well beyond quarter length of wing, sometimes at about half wing length; usually 11-20 presectoral crossveins (fig. 185) ▸	*Cosina* (not in Malesia?)
–	Hindwings with vein Rs arising at about a quarter wing length; usually 5-8 presectoral crossveins (fig. 186) ▸	3
3	Hypostigmatic cell in hindwings more than seven times as long as deep; hind margin regularly rounded (fig. 186) ▸	*Heoclisis*
–	Hypostigmatic cell in hindwings less than four times as long as deep, hind margin concave beyond hypostigmatic cell ▸	*Madrastra*

Notes on the genera of Acanthaclisinae

These are generally large (forewing length commonly up to 30-45 mm), sturdy and hairy antlions with dense venation, and the genera are separated predominantly on features of venation and legs, especially on claw and spur characteristics. Few species of each genus have been reported from the region, and the group is poorly represented.

Some genera have been characterised only poorly, but at least three of those currently recognised have been reported to occur in Malesia. One, *Heoclisis*, occurs also in Australia, but it is doubtful whether species reported from 'East Indies' as *Acanthaclisis* Rambur or *Stiphroneura* Gerstaecker truly represent those genera. Many species at some time attributed to *Acanthaclisis* are now *Heoclisis*, whilst others are now known to represent other major groups of antlions. *Acanthaclisis* is basically a Palaearctic genus. *Stiphroneura* includes two species, one from Africa, the other from 'East Indies' and reported also from India and Myanmar: with the attendant ambiguity of this label noted earlier, it has not yet been confirmed from Malesia. *S. inclusus* (Walker) (forewing length approximately 70 mm) is one of the largest antlions and is characterised by extensive forewing mottling (especially around the margin), with the hindwing having two large brown spots. A fifth genus, *Madrastra* Navás from 'Minduro' (presumed Mindoro, Philippines), is similar to *Heoclisis*, but differs in hindwing features.

Genitalic differences between genera are relatively small, and the abdominal apex is of the general form shown in fig. 189 (male) and fig. 190 (female).

Cosina Navás (type species *Acanthaclisis maclachlani* van der Weele). *Cosina* is distinguished by the distal origin of hindwing vein Rs. It is otherwise very similar to *Heoclisis*, but differs also in that the wings are usually narrower, and the tibial spurs somewhat shorter. Five Australian species are known.

Heoclisis Navás (type species *Myrmeleon fundatus* Walker). Most species of *Heoclisis* are Australian, but one possible inclusion is *Acanthaclisis fuscata* Fabricius from the 'East Indies'; its identity has not been confirmed. Likewise, the generic placement of a species from Malaysia needs confirmation. *H. louiseae* Banks keys here; it is excluded from *Acanthaclisis* by the form of the tibial spurs, but the hypostigmatic cell is relatively shorter than in many other species, being only about five times as long as deep (two syntypes examined). It occurs in West Malaysia.

Madrastra Navás (type species *Madrastra handlirschi* Navás). The sole included species is distinguished on generic features listed above. It is poorly known.

FIGURES 191 - 196
Myrmeleontidae - Myrmeleontini: wing venation of **191** *Myrmeleon*, **192** *Hagenomyia* — **193** abdominal apex, lateral aspect of male *Myrmeleon acer* and **194** of female *M. acer*; **195** idem of male *Hagenomyia* and **196** of female *Hagenomyia*.

191

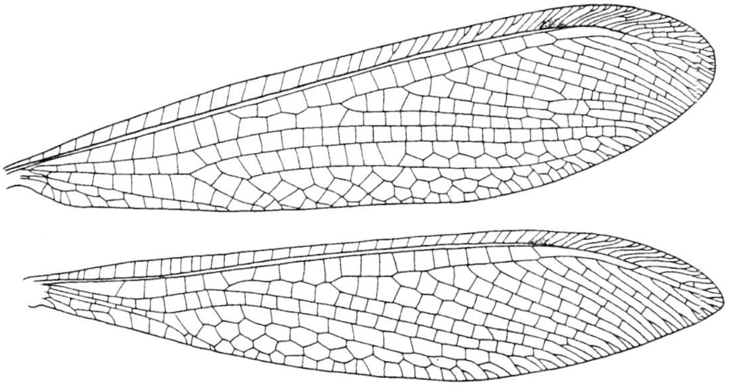

192

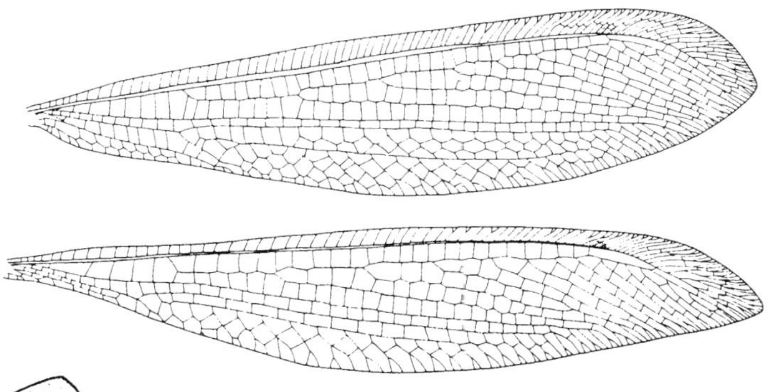

193

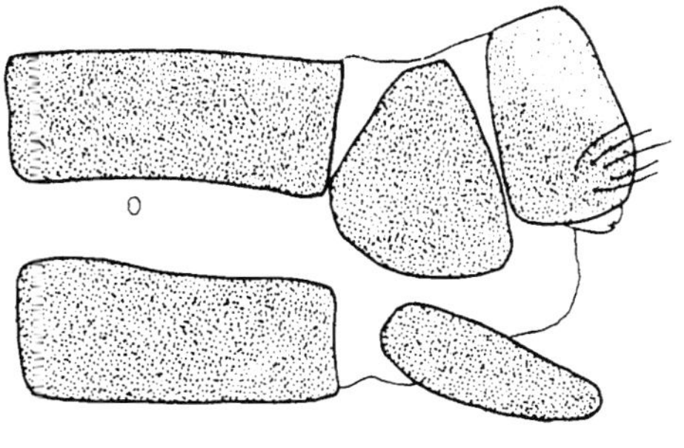

194

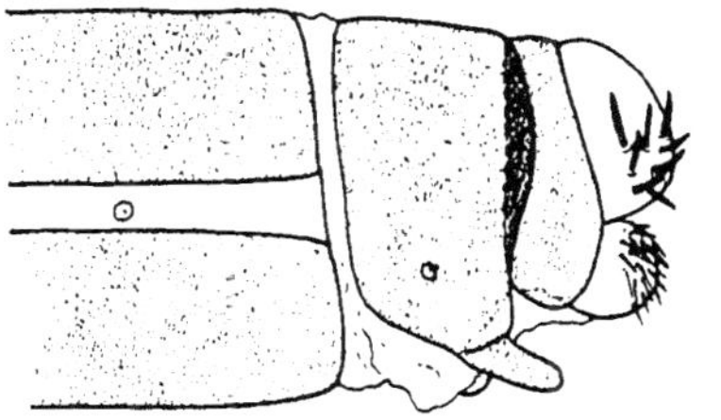

195

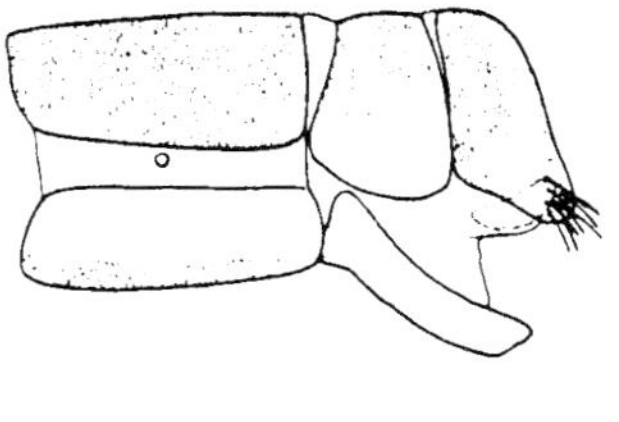

196

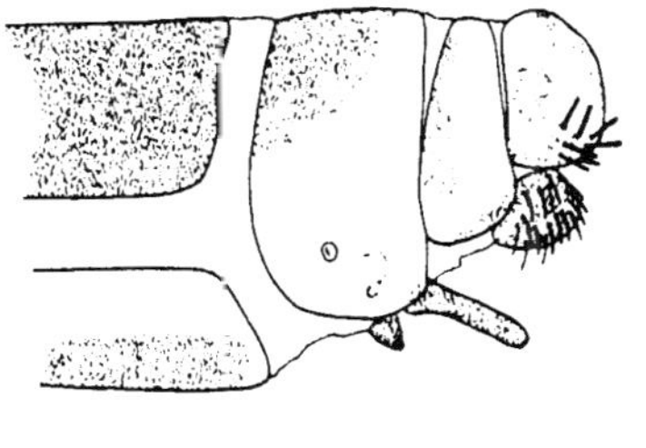

Key to the genera of Myrmeleontini

This tribe includes the typical antlions, whose larvae are the familiar pit-constructing species. It is widespread through the region and elsewhere.

1 Forewings with costal cells simple (fig. 191) ▸ *Myrmeleon*
– Forewings with few costal cells immediately basal to pterostigma biareolate (fig. 192) ▸ *Hagenomyia*

Notes on the genera of Myrmeleontini

Myrmeleon Linnaeus (type species *Myrmeleon formicarium* Linnaeus; figs. 191, 193, 194) is the predominant genus of this tribe and has a broad global distribution. The species are generally predominantly black-bodied with rather narrow wings. *M. pictifrons* Gerstaecker has broader wings, and more resembles *Hagenomyia*. The synonymy of this genus though is somewhat confused, and some workers continue to recognise a variety of other generic names (such as *Grocus* Navás, *Morter* Navás, *Leptoleon* Esben-Petersen, and others), mostly based on very small characters; in some ways then '*Myrmeleon*' is akin to the former use of '*Chrysopa*' as a 'holding genus' but with the proviso that it may indeed be the correct placement for most attributed species. The species from Malesia include several in the possible complex associated with *M. acer* Walker; *M. frontalis* Burmeister belongs to this group. *M. celebensis* MacLachlan, reported widely from Malaysia and Indonesia, is also a member of this complex and, although differentiated from *M. frontalis* by Banks on small venational features, the specific names have been applied somewhat uncritically to antlions in the region, and distribution of the various taxa is thus unclear.
The complexity of this group was indicated by Van der Weele (1909a) in his recognition of eleven subspecies of *M. acer* from the region, most of which seem to be only doubtfully distinct.

Hagenomyia Banks (type species *Myrmeleon tristis* Hagen; figs. 194, 197, 198) is a smaller genus, in which the species have broader wings than most *Myrmeleon* and a somewhat more complex venation. Taxonomy of species in Malesia is confused, and a number of different forms probably masquerade under the name *H. sagax* (Walker); there are apparently several forms with slightly different head markings, such as the pattern on the frons, but their genitalia are all very similar. Four regional subspecies were listed by Van der Weele (1909a). Although *Baliga* Navás is usually treated as a synonym of *Hagenomyia*, following

FIGURES 197 - 202

Myrmeleontidae - Protoplectrini, Distoleontini: *Pseudoformicaleo nubecula* - **197** wing venation (NOTE very long Cu2-fork in forewing) — **198** abdominal apex, lateral aspect of male genitalic complex, **199** abdominal apex of female and **200** of male — **201** wing venation of *Eophanes*, and **202** of *Bandidus*.

197

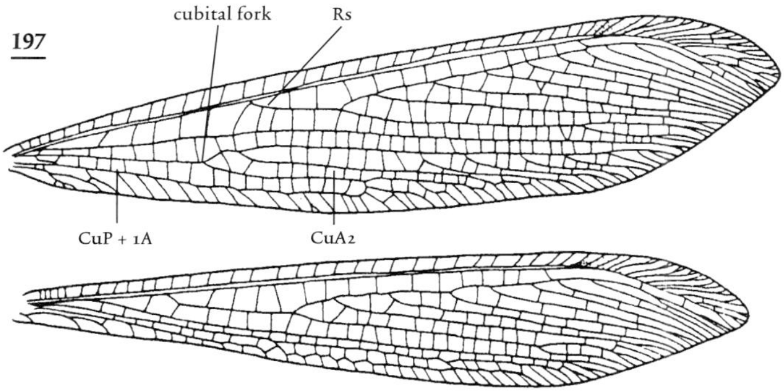

198

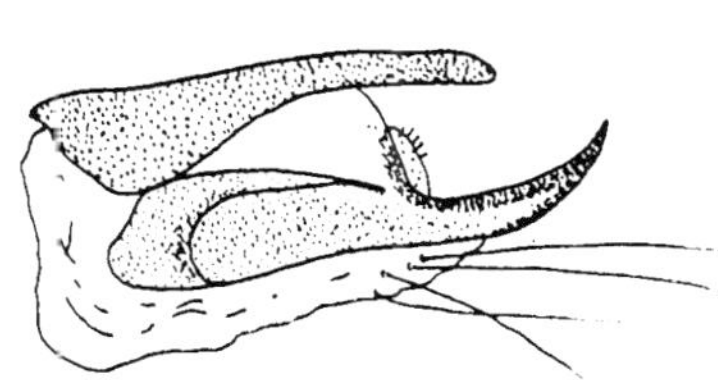

199

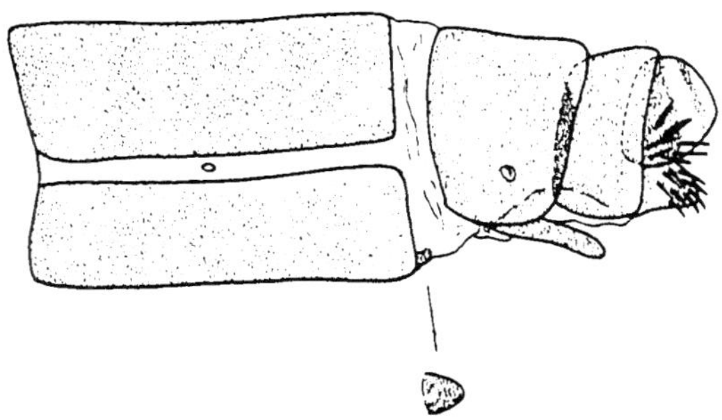

200

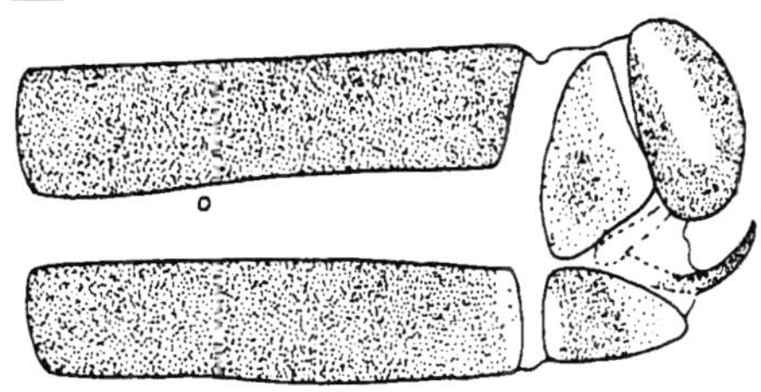

201

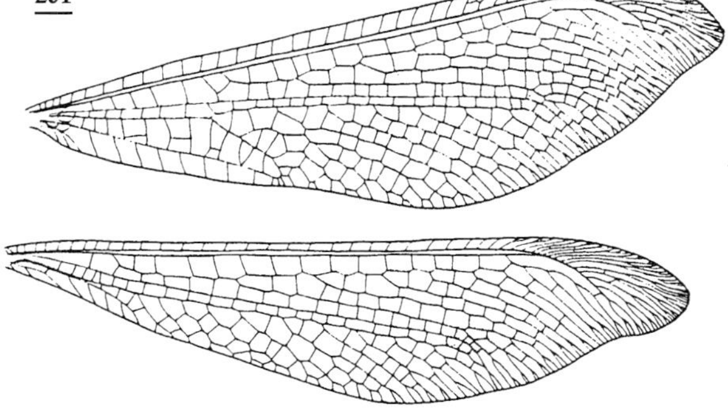

202

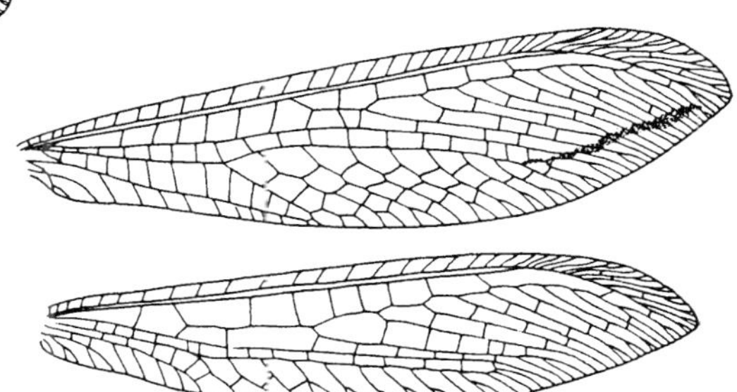

Markl (1954), this name was used for Taiwanese species by Stange & Wang (1998).

Neurocolinus Navás (type species *Colinus philippinus* Navás) can not be placed with certainty at the present. It was diagnosed (as *Colinus* Navás, a name later replaced by *Neurocolinus*) as 'Similis genero *Myrmeleon* L. rest.', and has several presectoral crossveins. It may be a synonym of *Myrmeleon*, but further investigation is needed.

Tribe Protoplectrini

Only one genus of this tribe has been reported from Malesia. It has very distinctive wing venation and is unlikely to be confused with other taxa in the region.

Pseudoformicaleo van der Weele (type species *Myrmeleon gracilis* Klug; figs. 197-200) has few species. The synonymy of *P. nubecula* (Gerstaecker) is complex, and the separate status of *P. jacobsoni* van der Weele (Indonesia) is not certain. *P. nubecula* is reported from Malaysia, Indonesia, Papua New Guinea, Australia, Sri Lanka, China, Taiwan and Japan, and is thus one of the most widely distributed species in the region. *Tahulus* Navás, *Gama* Navás and *Pseudoplectron* Navás are synonyms.

Tribe Distoleontini

Distoleontini is one of the predominant tribes of antlions in Malesia and, with Dendroleontini, contains many of the more distinctive species. The numerous species referred in literature to *Formicaleo* (or *Formicaleon*) belong in this tribe, but some can not be placed properly to genus. Banks (1911) raised *Distoleon* as a replacement name for *Formicaleo* Hagen, which has the same type species as *Myrmeleon*. For some members of the tribe, both taxonomic affiliation and nomenclatorial history remain confused. Study of the complex Australian fauna of Distoleontini (New, 1985b) showed that several of the features considered of generic value are, in fact, rather variable. Some 'residual' '*Formicaleo*' species are listed as such in the checklist, pending further study to determine their correct generic attributions. Most of these have been described only superficially.

Likewise, a number of other genera tentatively included in this group, following allocation by Markl (1954), can not be defined precisely at present. As used here '*Bandidus*' is a holding genus for a variety of species which may later be transferred elsewhere.

Key to the genera of Distoleontini

NOTE

This key does not include *Neuroleon* Navás. *Neuroleon* is predominantly an African genus, but four species referred to the genus have been reported from Malesia: *N. indistinctum* Navás and *N. uniformis* Navás from Malaysia, *N. disjunctus* Banks from the Philippines and East Malaysia, and *N. decoratum* Handschin from Timor. It is uncertain whether this allocation is correct: the species apparently are embraced within a broad concept of *Bandidus*, which has rather variable wing features, as implied above. True *Neuroleon* usually have a forewing posterior Banksian line and strongly hooked male parameres. The forewing of '*N.*' *decoratum* (fig. 203, after Handschin, 1937) does not have a posterior Banksian line, but this is clearly present in *N. disjunctus*.

In the key, t2, t3 and t4 are the second, third and fourth tarsal segments, respectively.

1 Wings very broad, falcate ▸ 2
– Wings narrow, not falcate ▸ 3

2 Both wings of similar shape (fig. 201) ▸ ***Eophanes***
– Hindwing considerably narrower than forewing ▸ ***Indoleon***

3 Tibial spurs absent ▸ ***Stenogymnocnemia***
– Tibial spurs present ▸ 4

4 Forewing vein Rs arising well before cubital fork (tibial spurs extending at least to apex of t3) ▸ ***Acratoleon*** Banks
– Forewing vein Rs arising almost opposite cubital fork (forewing posterior Banksian line absent, as in fig. 202; tibial spurs extend to t4) ▸ ***Bandidus*** (part)
– Forewing vein Rs arising distinctly beyond cubital fork (fig. 197) ▸ 5

5 Tibial spurs very long, extending to or beyond apex of t4 (fig. 204); male parameres fused and strongly hooked; forewing posterior Banksian line absent (figs. 205, 206); species usually large and sturdy ▸ ***Distoleon***
– Tibial spurs usually extend to t2, or are shorter; if longer, male parameres not elaborated or hooked (species usually slender; forewing posterior Banksian line present (fig. 1) or absent)
▸ (***Paraglenurus*** van der Weele also keys here) ***Bandidus*** (part)

FIGURES 203 - 209

Myrmeleontidae - Distoleontini: **203** forewing of *Neuroleon decoratum* (from Handschin, 1937) — **204** tibial spurs of *Distoleon bistrigatus* — **205** wing venation of *Distoleon somnolentus* — *Distoleon bistrigatus*, **206** wing venation, **207** abdominal apex of female, **208** abdominal apex of male, **209** male, genitalic complex.

203

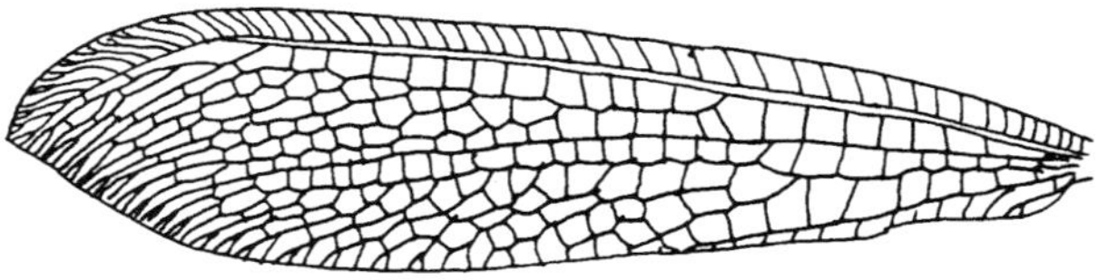

204

205

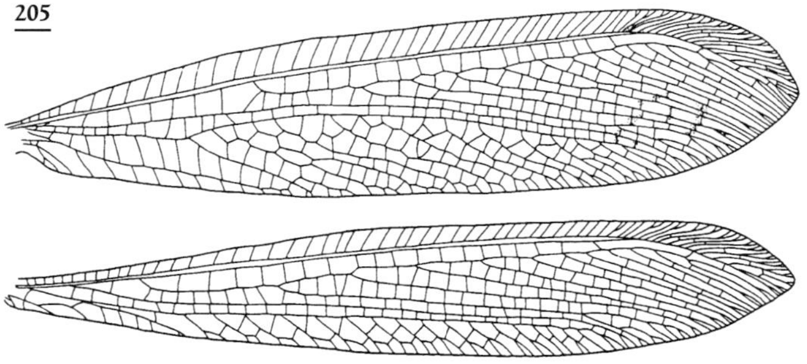

206

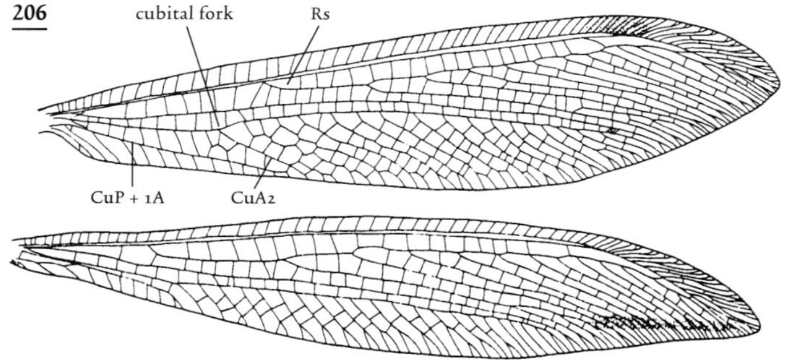

207

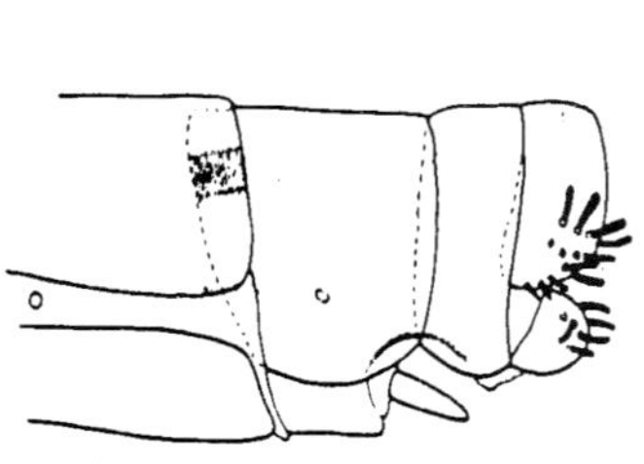

208

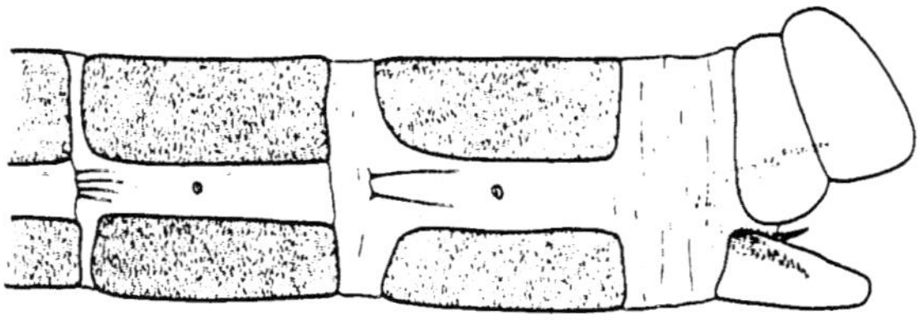

209

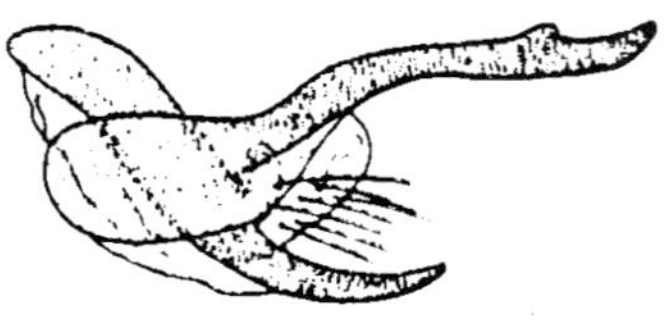

Notes on the genera of Distoleontini

Brisus Navás (type species *Brisus salomonicus* Navás) is very poorly known but is apparently a distoleontine. The sole described species, from the Solomon Islands, might occur elsewhere in eastern Malesia. Navás (1931) considered it related to *Neuroleon*, and no Banksian lines are present. It is not keyed above.

Acratoleon Banks (type species *Acratoleon flavum* Banks) is a poorly characterised genus with two regional species. *A. dispar* Banks has a trace of a wing 'bulla' (as in *Nuglerus*), although not as pronounced.

Distoleon Banks (type species *Distoleon verticalis* Banks) (figs. 204-209) is represented predominantly by two species, both rather variable in appearance. *D. bistrigatus* (Rambur) is widespread in the western Pacific region and Australia, and is large (forewing length to approximately 40 mm), with a prominent preapical dark streak on the strongly tapered hindwing. *D. somnolentus* (Gerstaecker) is smaller (forewing length usually less than 30 mm), lacks the dark hindwing streak, and the wing apices are more rounded; in the forewing, particularly, many outer gradate veins are 'shaded' with grey or brown. The latter species is, perhaps, the commonest non-*Myrmeleon* antlion over much of New Guinea and elsewhere, but the two species often occur in the same localities. *D. morpheus* (Kirby) from Christmas Island is now treated as a synonym of *D. somnolentus*. *Formicaleon* Banks and *Eidoleon* Esben-Petersen are synonyms.

Eophanes Banks (type species *Eophanes formosa* Banks) is very distinctive on its broad falcate wings and large size (cf. *Episalus*). *Eophanes formosa* has been recorded from the Philippines; as the only other member of the genus is Australian it could possibly be found in intervening countries.

Indoleon Banks (type species *Myrmeleon tacitus* Walker) has both wings slightly falcate, as in *Eophanes*, but the hindwing is distinctly narrower than the forewing and more acuminate. *I. tacitus* is a large species (forewing length approximately 50 mm). Both wings are marked strongly with dark brown. Note that '*Indoleon lacroixi* Fraser', purported to be from Indochina, was transferred to the New World brachynemurine genus *Hesperoleon* Banks by Markl (1954) and appears to have been misinterpreted geographically by Fraser.

Stenogymnocnemia Esben-Petersen (type species *Myrmeleon malus* Walker) includes several small and slender antlions with narrow wings, characterised by lack of tibial spurs. The species are elusive, perhaps rare, and all are known from very few specimens. Two Australian species are complemented by *S. ismayi* New, tentatively placed in this genus and known from Papua New Guinea.

Bandidus Navás (type species *Formicaleo canifrons* Navás) are rather sturdy antlions. The variety of their wing venation (especially the irregular incidence of the forewing posterior Banksian line) suggests that a complex might be represented. However, in the broad sense adopted here, *Bandidus* is particularly diverse in Australia, with a few additional species known from New Guinea (New, 1990a).
The more widespread and confusing representation centres on *B. vafer* (Walker) and its putative synonyms (Van der Weele 1909a, as *Formicaleo audax*). *Heteroleon* Esben-Petersen and *Formicaleo* sensu some authors, especially in relation to Australian fauna, are synonyms (New, 1985c).

Paraglenurus van der Weele (type species *Myrmeleon scopifer* Gerstaecker) keys to this complex (*Bandidus* s.l.) in the foregoing key. The hindwing is distinctly shorter than the forewing in *Paraglenurus* and the wings are generally of similar length in *Bandidus*. However, the two species of *Paraglenurus* are poorly documented and further clarification of generic limits is needed.

Tribe Dendroleontini

Dendroleontini is the most diverse group of Myrmeleontidae in the Australian region, and 83 Australian species have been described (New, 1985b). The Malesian fauna is relatively depauperate but contains taxa closely related to the Australian species and others which are very distinct. The world genera of the tribe were appraised by Stange (1976), and some changes have been made since then. As treated here, the tribe includes a few species of doubtful affinity. With the exception of two species of *Dendroleon* in North America, the tribe does not occur in the New World.

In the following key, *Cymothales* Gerstaecker is included on the basis of an unconfirmed record of this African genus from Borneo. *Phanoleon* Banks is not keyed separately (following Stange's (1976) implication that it is a synonym of *Dendroleon*), and *Nomes* Navás can not be included because it is very poorly characterised (New, 1990a). *Nomes* was considered a possible synonym of *Dendroleon* by Stange (1976). The single known specimen, badly damaged, appears to be more properly allied to *Mossega*, also considered by Stange (1976) to be a synonym of *Dendroleon*.

Glenurus Hagen is a wholly New World genus, but the name has been applied loosely elsewhere, including in combination for several species now placed in *Dendroleon*. In this context, the name has occasionally been employed for species which might occur in Malesia.

Key to the genera of Dendroleontini

1 Wing apex with a strongly convex 'bulla' in membrane; hypostigmatic cell in both wings with crossveins; basal tarsal segment usually much longer than distal segment (female: gonocoxite viii with posterior gonapophysis relatively small and cylindrical, anterior gonapophysis absent) ▸ *Nuglerus*
– Wing apex without bulla; hypostigmatic cell in both wings largely without crossveins (fig. 210); basal tarsal segment longer or shorter than distal tarsal segment ▸ 2

2 Hindwings very strongly tapered and longer than forewings; large species (forewing length to approximately 55 mm); tarsal segment 5 as long as segments 1-4 (female: posterior gonapophysis elongate; anterior gonapophysis absent) ▸ *Episalus*
– Hindwings not as above; smaller species ▸ 3

3 Hind margin of forewings strongly concave near apex; wings strongly patterned with brown/grey (female: posterior gonapophysis large, broad; anterior gonapophysis absent) ▸ *Periclystus*
– Hind margin of forewings not strongly concave near apex; wings not strongly patterned (female: anterior gonapophysis present, both gonapophyses digitiform) ▸ 4

4 Forewing vein Rs arising well beyond cubital fork; presectoral crossveins in forewings biareolate (hindwing longer than forewing; antennae weakly clubbed) ▸ *Cymothales* (in Malesia?)
– Forewing vein Rs arising opposite or, commonly, well before cubital fork; presectoral crossveins in forewings usually not biareolate, or only sporadically so (antennae distinctly clubbed) ▸ 5

5 Pretarsal claws opposable on last tarsal segments (fig. 218) ▸ *Mossega*
– Pretarsal claws not opposable on last tarsal segments ▸ 6

6 Female: gonocoxite viii with anterior gonapophysis twice as long as wide; male: sternite ix not divided ▸ *Nepsalus*
– Female: gonocoxite viii with anterior gonapophysis 1.5 or less times as long as wide (fig. 212); male: sternite ix divided into two lobes ▸ *Dendroleon*

FIGURES 210 - 212

Myrmeleontidae - Dendroleontini: wing venation of **210** *Episalus zephyrinus* (after New, 1990a), **211** *Periclystus vicinus*, **212** *Dendroleon kiungaensis* (after New, 1990a).

210

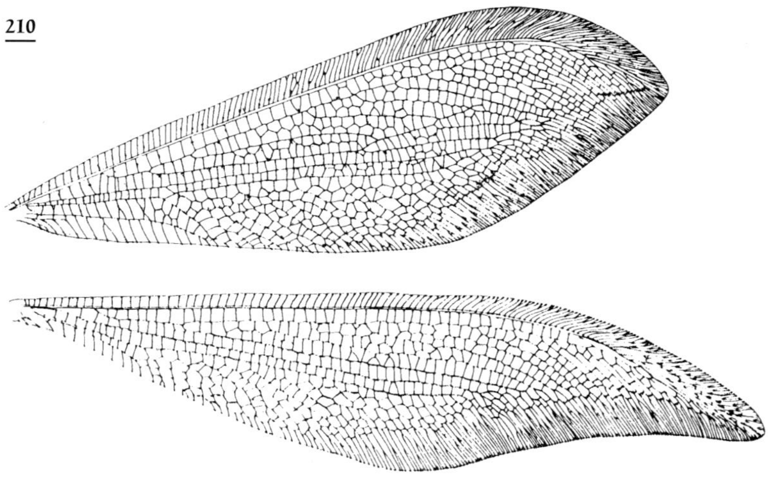

211

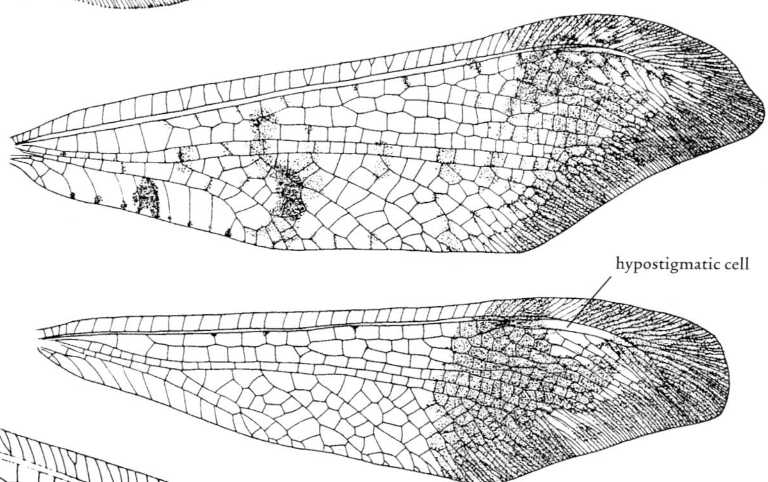

212

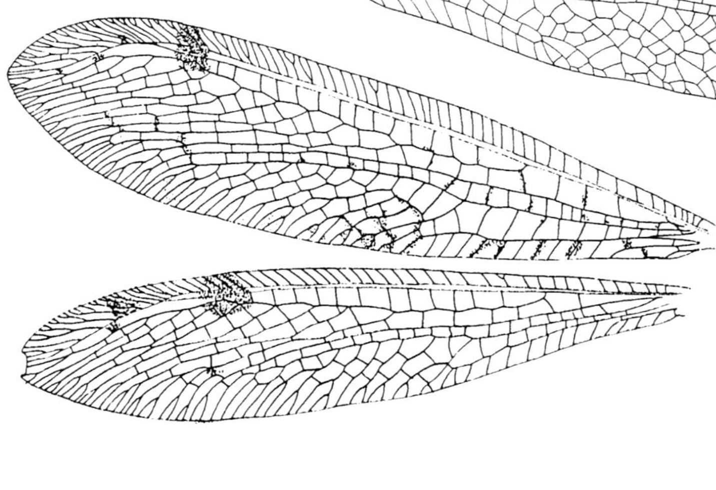

Notes to the genera of Dendroleontini

Dendroleon Brauer (type species *Myrmeleon pantherinum* Fabricius) is widely distributed in Australia, the Oriental and eastern Palaearctic regions, but many of the dozen or so regional species have not been defined critically. The types of several species (such as *D. javanus* (Banks)) are badly damaged, and these species may remain anomalous. A New Guinean species, *D. kiungaensis* New (fig. 212), appears to be closely related to Australian species.

D. vitripennis (Navás), described in *Neglurus*, was considered by Banks (1929) to be congeneric with *Nepsalus*, but Stange (1976) did not adopt that arrangement. The species has large dark spots on the wings. It is known from the Batu Caves, Malaysia, where larvae have also been found.

Neglurus Navás, *Borbon* Navás, *Madagascaroleon* Fraser are synonyms, but not *Mossega* Navás (which was also included by Stange (1976); see below).

Mossega Navás (type species *Mossega reticulata* Navás; figs. 213, 216) was synonymised with *Dendroleon* by Stange (1976) but reinstated by New (1985b). The two genera are closely related, but differ in (1) the form of the tarsal claws, as noted in the key, (2) the hindwing of *Mossega*, which is usually longer and narrower than the forewing, (3) the usually relatively longer tibial spurs of *Mossega*, and (4) the more basally originating vein Rs in both wings. *Bullanga* Navás is almost certainly a synonym of *Mossega*, but the type species has not been dissected; the type appears to be very similar to *Mossega* (New, 1985b). *Bullanga* was recognised by Stange (1976) and keyed on the characters used here to diagnose *Mossega*.

Mossega occurs in Australia, with two additional species described from New Guinea (New, 1990b). *Nomes lorianus* Navás is apparently similar to one of these two, but considerably smaller.

Nepsalus Navás (type species *Nepsalus indicus* Navás) is also difficult to distinguish, as the type of the type species is in poor condition: most of the abdomen is missing. Separation in the key on genitalic characters is after Stange (1976). *Nepsalus indicus* is from West Malaysia, and the only other included species, *N. insolitus* (Walker), is from India. *Nepsalus indicus* is a slender, delicate-looking antlion with very short antennae; both wings are tapered and much of the apical region of the wings is brown. Stange & Wang (1998) illustrated a presumed species of *Nepsalus* from Taiwan: it has strongly marked wings, the hindwing is conspicuously longer than the forewing and has a slightly falcate, tapered apex. The wing shape is distinctive, but no genitalic information was provided. See also comment under *Cymothales*.

FIGURES 213 - 218

Myrmeleontidae - Dendroleontini: **213** wing venation of *Mossega* — abdominal apex of females (lateral aspect) of **214** *Episalus*, **215** *Periclystus*, **216** *Mossega*, **217** *Dendroleon* — **218** *Mossega*, opposable pretarsal claws.

213

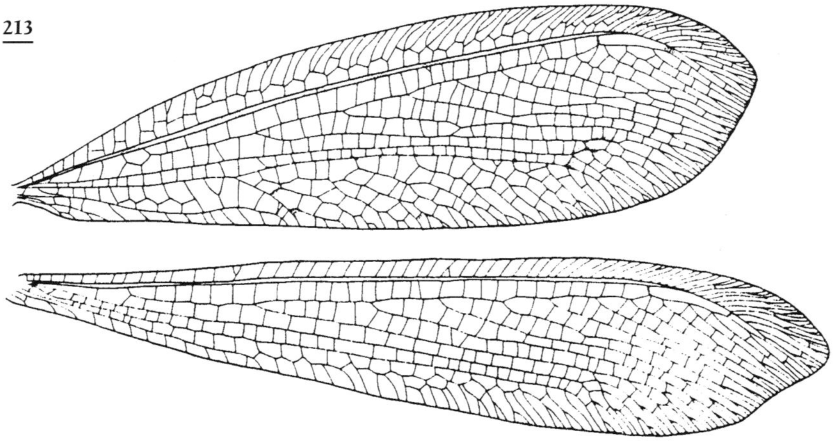

214

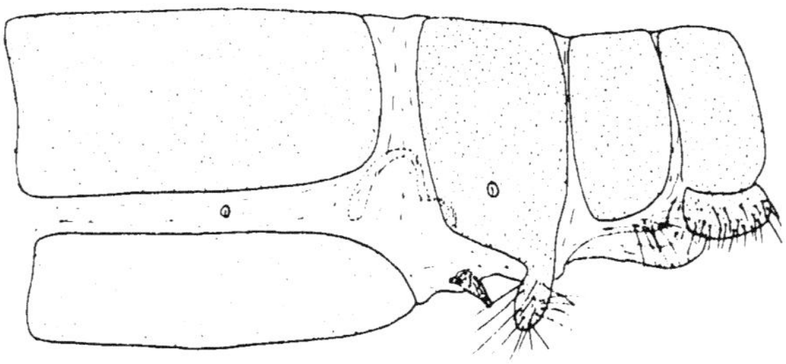

215

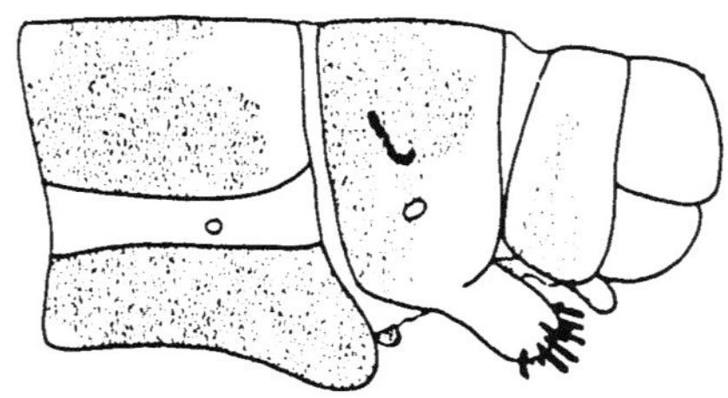

216

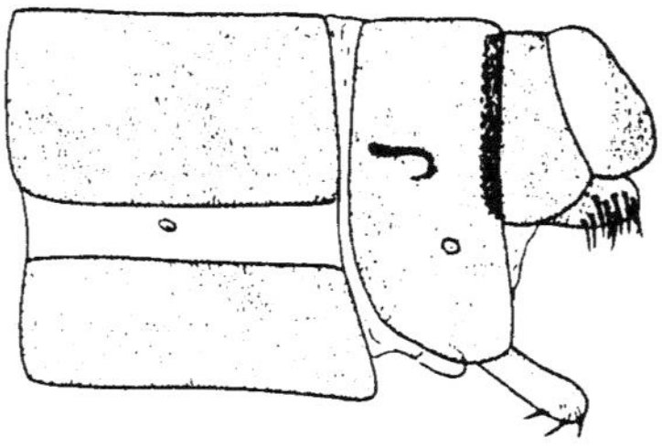

217

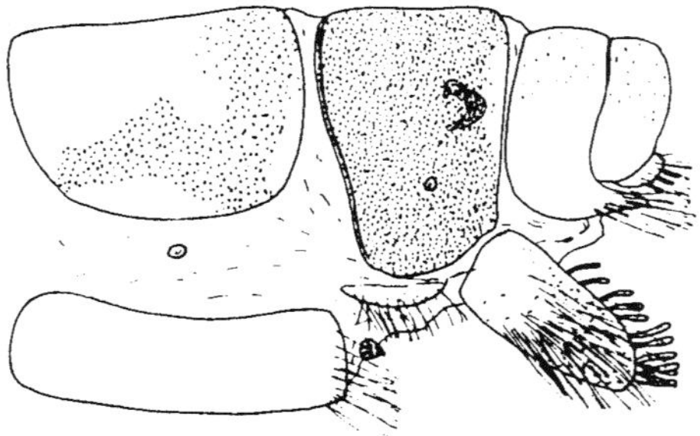

218

Episalus Gerstaecker (type species *Episalus zephyrinus* Gerstaecker) (figs. 210, 214) contains a single species which is among the most distinctive and spectacular of all antlions. It is also the largest species in Malesia. It may not be placed properly in Dendroleontini and shows some features allying it with Glenurini. *E. zephyrinus* is widespread in New Guinea and on some nearby islands, extending to the Solomons (Kimmins, 1957). It was redescribed by New (1990a).

Periclystus Gerstaecker (type species *Periclystus laceratus* Gerstaecker) contains three spectacularly marked species from Australia and one other, known from a single specimen, from Papua New Guinea (figs. 211, 215; New, 1985d). The appearance of the genus, with strongly excavated wings, is unmistakable in the region. The New Guinean species is clearly very closely related to the Australian taxa.

Nuglerus Navás (type species *Nuglerus scalaris* Navás) was placed by Stange (1976) in an independent subtribe, Nuglerina. The type species, *N. scalaris* Navás, occurs in Malaysia, but was described from India. *Bofia* Navás is a synonym and its type species (*Bofia maculata* Navás) also occurs in West Malaysia. It is very similar in appearance to *N. scalaris* but considerably smaller (forewing length about 22 mm, compared with about 33 mm in *N. scalaris*). The raised bulla in the pterostigmal region of both wings is very distinctive.

Cymothales Gerstaecker (type species *Cymothales mirabilis* Gerstaecker) is almost certainly recorded erroneously from Malesia, and none has been reported in recent years. In his revision of the genus, Mansell (1987) recognised 17 species, all from tropical Africa, and excluded '*C.*' *lestagei* Lacroix without reallocation to any other genus. Mansell (1987) noted that the holotype of this taxon from Borneo is in poor condition, so that more material is needed to clarify its affinities. An entire specimen labelled as *C. lestagei* in the Natural History Museum, London (forewing length 42 mm, body length approximately 46 mm) is placed there under *Nepsalus*. Any similar specimens from Malesia will be of considerable value in clarifying the generic status of this anomalous taxon. See also the comment under *Nepsalus*.

Biological outline

Although antlions are best-known for the pit-forming habit of larvae, this habit is in fact very restricted within the family and probably only *Myrmeleon* and *Hagenomyia* larvae form pits in Malesia. Larvae of many other genera are more general 'lurkers' or more active foragers in sand or soft soil, on or under rock shelves, on bark or under logs. Indeed, Mansell (1999) attributed the success of the family to their wide variety of larval habits related to psammophily. The life cycle of some species takes at least 2-3 years and is thus much longer than for most non-myrmeleontoid Neuroptera. Eggs are apparently scattered over the substrate. Pupation takes place in sand.

MAJOR REFERENCES — Wheeler (1930): general account of larval biology; Markl (1954): world synopsis of genera, with comments on relationships; Hölzel (1972): synopsis of eastern palaearctic taxa; Stange (1976): synopsis of genera of Dendroleontini; New (1985a, b, c): revision of Australian taxa; Mansell (1985, 1990): discussion of Palparini; Hölzel (1986): zoogeography.

chapter 5

Checklist of Neuroptera from Malesia

In this checklist lacewings described or recorded from Malesia and some nearby parts of Southeast Asia are included. For each listed species, a reference to the original description is given, and all available records are listed by country. Ranges outside Malesia for species recorded from Malesia are given in square brackets. For comparison, the fauna of Malesian families reported from Taiwan is included as the most fully documented fauna in Southeast Asia. These species, and other species known only from outside Malesia, are indicated by asterisks, and are listed simply to indicate possible additional taxa not yet found within Malesia. However, the complex fauna of Australia is not listed other than for Christmas Island which, biogeographically, is part of Indonesia; it is catalogued by New (1996). Synonymy is not given, although a few uncertain synonyms are indicated in parentheses.

Precise political attribution of distributional records from Malesia is sometimes difficult - for example, with early records from 'New Guinea', 'Borneo', or 'Malaysia'. Nowadays 'New Guinea' refers to the whole island of New Guinea and its immediate neighbours. It thus comprises Papua (Irian Jaya) (the western half of the island, a province of Indonesia) and Papua New Guinea (the eastern half of the island). Likewise, the Solomon Islands have been independent for more than 20 years but were formerly part of Papua New Guinea. Their fauna is closely allied to the Papua New Guinea species. 'Borneo' refers to any part of the island now comprising Sarawak, Sabah, Brunei and the Indonesian province of Kalimantan. Most older records of 'Malaysia' refer to West (peninsular) Malaysia, although taxa from Singapore and East Malaysia are also included under this broad heading. 'East Malaysia' includes Sabah and Sarawak, but there are no specified records from Brunei. Confusion over the attribution 'East Indies' was discussed earlier under Rapismatidae and Myrmeleontidae.

NOTE

[ ] range outside Malesia, but also known from Malesia

* known only outside Malesia

Rapismatidae

Rapisma MacLachlan, 1866

R. *barnardi* New, 1985. Neuroptera International 3: 133 E. Malaysia: Sabah

R. *berhalense* Barnard & New, 1985. Neuroptera International 3: 168 ▸ Indonesia: Malacca Strait, Berhala Island

*R. *burmanum* Navás, 1929. Annali del Museo civico di storia naturale di Genova 53: 377 ▸ [Myanmar]

R. *corundum* Barnard, 1981. Systematic Entomology 6: 132 ▸ W. Malaysia

R. *cryptunum* Barnard & New, 1985. Neuroptera International 3: 168 ▸ W. Malaysia

R. *haesitum* Barnard & New, 1985. Neuroptera International 3: 169 ▸ Indonesia: Java

*R. *intanonum* Barnard, 1981. Systematic Entomology 6: 134 ▸ [northern Thailand]

R. *johorense* Barnard & New, 1985. Neuroptera International 3: 169 ▸ W. Malaysia

R. *malayanum* Barnard, 1981. Systematic Entomology 6: 132 ▸ W. Malaysia

R. *propinquum* Barnard & New, 1985. Neuroptera International 3: 169 ▸ W. Malaysia

R. *sabahnum* Barnard, 1981. Systematic Entomology 6: 134 E. Malaysia: Sabah

Coniopterygidae

ALEUROPTERYGINAE

Coniocompsa Enderlein, 1905

*C. *furcata* Banks, 1937. The Philippine journal of science 62: 277 ▸ [Taiwan]

C. *meinanderi* Monserrat, 1982. Annali del Museo civico di storia naturale di Genova 84: 12 ▸ Malaysia; Singapore; Indonesia

C. *mindanaoensis* Meinander, 1972. Acta zoologica Fennica 136: 98 ▸ Philippines

C. *traceyae* New, 1990. Neuroptera International 6: 2 ▸ E. Malaysia: Sabah

C. *vesiculigera* Enderlein, 1905. Zoologischer Anzeiger 29: 225 ▸ Malaysia

Cryptoscenea Enderlein, 1914

C. *novaeguineensis* Meinander, 1972. Acta zoologica Fennica 136: 109 ▸ Papua New Guinea

NOTE

One species has been recorded simply as '*Cryptoscenea* sp.':

'*Cryptoscenea* sp.' Monserrat, 1982. Annali del Museo civico di storia naturale di Genova 84: 11 ▸ Indonesia: Bali

Heteroconis Enderlein, 1905

H. allisoni New, 1988. Neuroptera International 5: 4 ▸ Papua New Guinea

H. (*Drepanoconis*) *amoena* Tjeder, 1973. Entomologisk Tidskrift 93: 204 ▸ Papua New Guinea

H. angustipennis Monserrat, 1982. Annali del Museo civico di storia naturale di Genova 84: 22 ▸ Indonesia

H. axeli New, 1988 (In: New & Sudarman 1988). Philosophical Transactions of the Royal Society of London B 322: 416 ▸ Indonesia: Krakatau Islands

H. bifurcata Meinander, 1990. Acta zoologica Fennica 189: 11 ▸ Papua New Guinea

H. candida Tjeder, 1973. Entomologisk Tidskrift 93: 201 ▸ Papua (Irian Jaya)

H. cornuta Monserrat, 1982. Annali del Museo civico di storia naturale di Genova 84: 20 ▸ Indonesia

H. curvata Meinander, 1990. Acta zoologica Fennica 189: 12 ▸ Papua New Guinea

H. dahli Enderlein, 1906. Zoologische Jahrbücher - Abteilung für allgemeine Zoologie und Physiologie der Tiere 23: 228 ▸ Papua New Guinea

H. enarotadiensis New, 1990. Invertebrate Taxonomy 4: 677 ▸ Papua (Irian Jaya)

H. fenestrata New, 1990. Invertebrate Taxonomy 4: 681 ▸ Papua New Guinea

H. flavicornuta Tjeder, 1973. Entomologisk Tidskrift 93: 188 ▸ Papua (Irian Jaya)

H. fumipennis Tjeder, 1973. Entomologisk Tidskrift 93: 198 ▸ Papua (Irian Jaya)

H. fusca Meinander, 1972. Acta zoologica Fennica 136: 85 ▸ Malaysia (W. Malaysia, Sabah)

H. gagnei New, 1988. Neuroptera International 5: 6 ▸ Papua New Guinea

H. interrupta (Banks, 1937). The Philippine journal of science 63: 153 ▸ Philippines

H. iriana Tjeder, 1973. Entomologisk Tidskrift 93: 194 ▸ Papua (Irian Jaya)

H. javanica Monserrat, 1982. Annali del Museo civico di storia naturale di Genova 84: 14 ▸ Indonesia

H. kaindiensis New, 1990. Invertebrate Taxonomy 4: 676 ▸ Papua New Guinea

H. kaitensis New, 1990. Invertebrate Taxonomy 4: 670 ▸ Papua New Guinea

H. madangensis Meinander, 1990. Acta zoologica Fennica 189: 14 ▸ Papua New Guinea

**H. nigripalpis* Meinander, 1972. Acta zoologica Fennica 136: 83 ▸ [Taiwan]

H. pennyi Meinander, 1990. Acta zoologica Fennica 189: 16 ▸ Papua New Guinea

H. pepa Monserrat, 1982. Annali del Museo civico di storia naturale di Genova 84: 17 ▸ Indonesia

H. (*Drepanoconis*) *pioraensis* New, 1990. Invertebrate Taxonomy 4: 667 ▸ Papua New Guinea

H. spinosa New, 1990. Invertebrate Taxonomy 4: 667 ▸ Papua New Guinea

H. striata New, 1990. Invertebrate Taxonomy 4: 679 ▸ Papua New Guinea

H. toxopei Tjeder, 1973. Entomologisk Tidskrift 93: 191 ▸ Papua (Irian Jaya)
H. umbrata New, 1990. Invertebrate Taxonomy 4: 668 ▸ Papua New Guinea
**H. vietnamensis* Meinander, 1990. Acta zoologica Fennica 189: 20 ▸ [Vietnam]
H. wauensis New, 1988. Neuroptera International 5: 8 ▸ Papua New Guinea
H. wilhelmensis New, 1990. Invertebrate Taxonomy 4: 675 ▸ Papua New Guinea

NOTE

Three species have been reported simply as '*Heteroconis* sp.':

'*Heteroconis* sp.' Monserrat, 1982. Annali del Museo Civico di Storia Naturale Giacomo Doria 84: 24 ▸ Indonesia
'*Heteroconis* sp.' New, 1990. Neuroptera International 6: 5 (two species) ▸ E. Malaysia: Sabah
'*Heteroconis* sp.' New, 1990. Invertebrate Taxonomy 4: 673, 674 (two species) ▸ New Guinea

Paraconis Meinander, 1972
P. borneoensis Meinander, 1972. Acta zoologica Fennica 136: 114 ▸ E. Malaysia: Sabah

Spiloconis Enderlein, 1907
S. notata (Navás, 1926). Memorie della Pontificia Accademia delle Scienze, Nuovi Lincei 2 (9): 119 ▸ Philippines
**S. sexguttata* Enderlein, 1907. Stettiner entomologische Zeitung 68: 7 ▸ [Thailand, Taiwan, China, Japan]

CONIOPTERYGINAE

Coniopteryx Curtis, 1834
C. javana Enderlein, 1907 (no subgenus allocated). Notes from the Leyden Museum 28: 224 ▸ Indonesia: Java
C. ralumensis Enderlein, 1906 (no subgenus allocated). Zoologische Jahrbücher - Abteilung für allgemeine Zoologie und Physiologie der Tiere 23: 203 ▸ Papua New Guinea
C. (Xeroconiopteryx) papuensis Meinander, 1990. Acta zoologica Fennica 89: 39 ▸ Papua New Guinea
C. (Coniopteryx) ambigua Withycombe, 1925. Memoirs of the Department of Agriculture in India / Entomological series 9: 13 ▸ Malaysia [China, India, Sri Lanka]
C. (C.) biroi Enderlein, 1906. Zoologische Jahrbücher - Abteilung für allgemeine Zoologie und Physiologie der Tiere 23: 203 ▸ Papua New Guinea

C. (C.) *exigua* Withycombe, 1925. Memoirs of the Department of Agriculture in India / Entomological series 9: 12 ▸ Malaysia: Sabah [India, Nepal, Pakistan]
C. (C.) *falciger* Karny, 1923. Treubia 3: 384 ▸ Singapore; Indonesia: Java
C. (C.) *goniocera* Meinander, 1972. Acta zoologica Fennica 136: 264
▸ Malaysia: Sabah [India, Sri Lanka]
*C. (C.) *insularis* Meinander, 1972. Acta zoologica Fennica 136: 243
▸ [Taiwan, Okinawa]
C. (C.) *kaindiensis* New, 1988. Neuroptera International 5: 10 ▸ Papua New Guinea
C. (C.) *lobifrons* Murphy & Lee, 1971. Journal of Entomology, series B 40: 152
▸ Singapore
C. (C.) *macroscapes* Meinander, 1990. Acta zoologica Fennica 189: 56
▸ Malaysia: Sabah
C. (C.) *morobensis* Meinander, 1990. Acta zoologica Fennica 189: 58
▸ Papua New Guinea
*C. (C.) *pallescens* Meinander, 1972. Acta zoologica Fennica 136: 242 ▸ [Taiwan]
C. (C.) *prehensilis* Murphy & Lee, 1971. Journal of Entomology, series B 40: 155
▸ Singapore, [China, India]
C. (C.) *tagalica* (Banks, 1937). The Philippine journal of science 63: 153
▸ Malaysia; Indonesia: Bali, Krakatau Islands; Philippines

Conwentzia Enderlein, 1905
C. *nietoi* Monserrat, 1982. Annali del Museo civico di storia naturale di Genova 84: 24 ▸ Indonesia, [Sri Lanka]

Neosemidalis Enderlein, 1930
N. (*Neosemidalis*) *antennalis* New, 1988. Neuroptera International 5: 10
▸ Papua New Guinea
N. (N.) *brevipennis* Meinander, 1990. Acta zoologica Fennica 189: 33
▸ Papua New Guinea

Semidalis Enderlein, 1905
S. *aleyrodiformis* (Stephens, 1836). Illustrations of British Entomology. Mandibulata VI. London: 116
▸ Malaysia; Thailand, [widespread over much of Palaearctic region]
S. *decipiens* (Roepke, 1916). Zoölogische Mededelingen 2: 157
▸ Malaysia: Sabah; Indonesia, [India]
S. *galantei* Monserrat, 1982. Annali del Museo civico di storia naturale di Genova 84: 28 ▸ Philippines
*S. *macleodi* Meinander, 1972. Acta zoologica Fennica 136: 314 ▸ [Taiwan]
S. *unicornis* Meinander, 1972. Acta zoologica Fennica 136: 330
▸ Malaysia [Taiwan, Mongolia]

Sisyridae

Sisyra Burmeister, 1839
S. indica Needham, 1909. Records of the Indian Museum 3: 206 ▸ Philippines; W. Malaysia; Thailand [India]
S. maculata Monserrat, 1981. Eos 57: 177 ▸ Thailand
S. bakeri Banks, 1913. Transactions of the American Entomological Society 39: 215 ▸ Philippines

Osmylidae

SPILOSMYLINAE

Spilosmylus Kolbe, 1897
**S. alleni* New, 1991. Invertebrate Taxonomy 5: 16 ▸ [Thailand]
S. alticolus Banks, 1937. The Philippine journal of science 63: 151 ▸ Philippines
S. amarillus New, 1988. Invertebrate Taxonomy 2: 812 ▸ Papua (Irian Jaya)
S. analis New, 1988. Invertebrate Taxonomy 2: 815 ▸ Papua (Irian Jaya)
S. apoanus Banks, 1937. The Philippine journal of science 63: 152 ▸ Philippines
S. araucariensis New, 1988. Invertebrate Taxonomy 2: 814 ▸ Papua (Irian Jaya)
S. ateritriangulus New, 1986. Australian Journal of Zoology 34: 870 ▸ Papua New Guinea
S. basalis New, 1986. Australian Journal of Zoology 34: 881 ▸ Papua New Guinea
S. bernhardensis New, 1988. Invertebrate Taxonomy 2: 818 ▸ Papua (Irian Jaya)
S. bimaculatus New, 1986. Australian Journal of Zoology 34: 857 ▸ Papua New Guinea
S. brandti New, 1986. Australian Journal of Zoology 34: 881 ▸ Papua New Guinea
S. conspersus (Walker, 1853). Catalogue of the specimens of Neuropterous insects in the collection of the British Museum II: 234 ▸ 'E. Indies'
S. cornutus New, 1988. Invertebrate Taxonomy 2: 814 ▸ Papua (Irian Jaya)
S. elongatus New, 1988. Invertebrate Taxonomy 2: 815 ▸ Papua (Irian Jaya)
S. formosus Banks, 1924. Bulletin of the Museum of Comparative Zoology at Harvard 65: 430 ▸ Philippines
S. fraternus (Banks, 1931). Journal of the Federated Malay States Museums 16: 417 ▸ E. Malaysia: Sabah
S. fumosus New, 1988. Invertebrate Taxonomy 2: 809 ▸ Papua (Irian Jaya)
S. fuscomaculatus New, 1988. Invertebrate Taxonomy 2: 816 ▸ Papua (Irian Jaya)
S. gressitti New, 1986. Australian Journal of Zoology 34: 864 ▸ Papua New Guinea
S. griseofasciatus New, 1986. Australian Journal of Zoology 34: 871 ▸ Papua New Guinea

S. inquinatus MacLachlan, 1870. The Entomologist's monthly magazine 6: 200 ▸ Indonesia: Ceram; Philippines

**S. inthanonensis* New, 1991. Invertebrate Taxonomy 5: 18 ▸ [Thailand]

S. ismayi New, 1986. Australian Journal of Zoology 34: 863 ▸ Papua New Guinea

S. javensis New, 1991. Invertebrate Taxonomy 5: 19 ▸ Indonesia: Java

S. latipennatus New, 1986. Australian Journal of Zoology 34: 872 ▸ Papua New Guinea

S. leletensis New, 1986. Australian Journal of Zoology 34: 870 ▸ Papua New Guinea

S. lieftincki New, 1988. Invertebrate Taxonomy 2: 817 ▸ Papua (Irian Jaya)

S. majalis Navás, 1924. Revista de la Academia de ciencias exactas, físicas químicas y naturales de Zaragoza 9: 32 ▸ Papua (Irian Jaya)

S. misoolensis New, 1988. Invertebrate Taxonomy 2: 812 ▸ Papua (Irian Jaya)

S. modestus (Gerstaecker, 1893). Mitteilungen aus dem Naturwissenschaftlichen Verein für Neu-Vorpommern und Rügen in Greifswald 25: 169 ▸ E. Malaysia: Sarawak; Indonesia

S. monticolus (Banks, 1937). The Philippine journal of science 63: 149 ▸ Philippines

S. navasi New, 1988. Invertebrate Taxonomy 2: 817 ▸ Papua (Irian Jaya)

S. nebulosus New, 1988. Invertebrate Taxonomy 2: 807 ▸ Papua (Irian Jaya)

S. nephelius Navás, 1926. Memorie della Pontificia Accademia delle Scienze, Nuovi Lincei (2) 9: 115 ▸ Philippines

S. neobritannicus New, 1986. Australian Journal of Zoology 34: 864 ▸ Papua New Guinea

S. obliquus New, 1986. Australian Journal of Zoology 34: 864 ▸ Papua New Guinea

S. ocellatus (Krüger, 1914). Stettiner entomologische Zeitung 75: 32 ▸ Indonesia: Java

S. pictus New, 1988. Invertebrate Taxonomy 2: 810 ▸ Papua (Irian Jaya)

S. pretiosus (Banks, 1931). Journal of the Federated Malay States Museums 16: 417 ▸ E. Malaysia: Sabah [India]

S. proximus Banks, 1937. The Philippine journal of science 63: 151 ▸ Philippines

S. punctatus New, 1988. Invertebrate Taxonomy 2: 811 ▸ Papua (Irian Jaya)

S. rattanensis New, 1988. Invertebrate Taxonomy 2: 805 ▸ Papua (Irian Jaya)

S. reflexus New, 1988. Invertebrate Taxonomy 2: 811 ▸ Papua (Irian Jaya)

S. sibilensis New, 1988. Invertebrate Taxonomy 2: 811 ▸ Papua (Irian Jaya)

S. sigiensis New, 1988. Invertebrate Taxonomy 2: 813 ▸ Papua (Irian Jaya)

S. sinuosus New, 1988. Invertebrate Taxonomy 2: 816 ▸ Papua (Irian Jaya)

S. subtilis New, 1986. Australian Journal of Zoology 34: 882 ▸ Papua New Guinea

S. sumatranus Krüger, 1913. Stettiner entomologische Zeitung 74: 25 ▸ Indonesia: Sumatra

S. sumbanus Krüger, 1913. Stettiner entomologische Zeitung 74: 25
▸ Indonesia: Sumba
S. tjederi New, 1988. Invertebrate Taxonomy 2: 810 ▸ Papua (Irian Jaya)
S. toxopeusi New, 1988. Invertebrate Taxonomy 2: 809 ▸ Papua (Irian Jaya)
S. tuberculatus Walker, 1853. Catalogue of the specimens of Neuropterous insects in the collection of the British Museum II: 235 ▸ 'E. Indies'
S. xavieri Navás, 1928. Boletín de la Sociedad Entomológica de España 11: 90
▸ Papua New Guinea
S. weiensis New, 1986. Australian Journal of Zoology 34: 869
▸ Papua New Guinea

Thaumatosmylus Krüger, 1913
T. delicatus (Banks, 1931). Journal of the Federated Malay States Museums 16: 418
▸ E. Malaysia: Sabah
T. diaphanus (Gerstaecker, 1893). Mitteilungen aus dem Naturwissenschaftlichen Verein für Neu-Vorpommern und Rügen in Greifswald 25: 168
▸ Indonesia: Java
**T. ornatus* Nakahara, 1955. Kontyû 23: 10 ▸ [Taiwan]
T. raongensis New, 1991. Invertebrate Taxonomy 5: 10 ▸ Indonesia: Java
T. umbratus New, 1991. Invertebrate Taxonomy 5: 8 ▸ E. Malaysia: Sabah

Mantispidae

MANTISPINAE

Asperala Lambkin, 1986
A. erythraea (Brauer, 1867). Verhandlungen der Zoologisch-Botanischen Gesellschaft in Wien 17: 506 ▸ Indonesia [Australia]

Austroclimaciella Handschin, 1961
**A. habutsuella* (Okamoto, 1911). Zoologischer Anzeiger 37: 300
▸ [Taiwan; China; Japan]
A. luzonica (van der Weele, 1909). Notes from the Leyden Museum 31: 94
▸ Philippines
A. quadrituberculata (Westwood, 1852). The Transactions of the Entomological Society of London (2) 1: 264 ▸ Indonesia: Java [Taiwan; India]
A. weelei Handschin, 1961. Nova Guinea Zoology 1: 289 ▸ Indonesia

Austromantispa Esben-Petersen, 1917

A. imbecilla (Gerstaecker, 1885). Mitteilungen aus dem Naturwissenschaftlichen Verein für Neu-Vorpommern und Rügen in Greifswald 16: 41
‣ Indonesia: Timor; Papua New Guinea [Australia]

Campanacella Handschin, 1961

C. hamiltonella (Westwood, 1867). The Transactions of the Entomological Society of London (1867): 506 ‣ Malaysia; Indonesia: Java

Climaciella Enderlein, 1910

C. ferrosa Navás, 1914. Memorias de la Real Academia de Ciencias exactas, físicas y naturales de Madrid 12: 482 ‣ Malaysia

*C. *leopoldi* Lestage, 1934. Bulletin et annales de la Société Entomologique de Belgique 74: 154 ‣ [Vietnam]

Entanoneura Enderlein, 1910

E. albertisii Navás, 1929. Annali del Museo civico di storia naturale di Genova 53: 381 ‣ Indonesia: Timor

*E. *feae* Navás, 1929. Annali del Museo civico di storia naturale di Genova 53: 382
‣ [Myanmar]

Euclimacia Enderlein, 1910

*E. *badia* Okamoto, 1910. Dobutsugaku Zasshi Tokyo 22: 543 ‣ [Taiwan]

E. basiflava Handschin, 1961. Nova Guinea Zoology 1: 273 ‣ Malaysia

E. celebica Handschin, 1961. Nova Guinea Zoology 1: 267 ‣ Indonesia: Sulawesi

E. ferdinandi Navás, 1928. Boletín de la Sociedad Entomológica de España 11: 137
‣ Borneo

E. flavicauda Esben-Petersen, 1917. Arkiv för zoologi 11: 7 ‣ Indonesia: Sumatra

E. flavocincta Stitz, 1913. Mitteilungen aus dem Zoologischen Museum in Berlin 7: 42 ‣ Solomon Islands

*E. *fusca* Stitz, 1913. Mitteilungen aus dem Zoologischen Museum in Berlin 7: 41 ‣ [Taiwan]

E. gerstaeckeri Banks, 1920. Bulletin of the Museum of Comparative Zoology at Harvard 64: 341 ‣ Singapore

E. guerini (Westwood, 1852). The Transactions of the Entomological Society of London (2) 1: 255 ‣ Indonesia: Moluccas, Ambon

E. jacobsoni Handschin, 1961. Nova Guinea Zoology 1: 262 ‣ Indonesia: Sumatra

E. metallica Esben-Petersen, 1917. Arkiv för zoologi 11: 10
‣ Indonesia: Sumatra, Sulawesi

E. morosa (Gerstaecker, 1893). Mitteilungen aus dem Naturwissenschaftlichen Verein für Neu-Vorpommern und Rügen in Greifswald 25: 149 ▸ Philippines: Palawan; Borneo

E. nigra Handschin, 1961. Nova Guinea Zoology 1: 274 ▸ Indonesia: Java

E. partita Enderlein, 1910. Stettiner entomologische Zeitung 29: 366 ▸ Indonesia: Sulawesi

E. regina Esben-Petersen, 1917. Arkiv för zoologi 11: 8 ▸ Singapore

**E. rhombica* Navás, 1914. Memorias de la Real Academia de Ciencias y Artes de Barcelona 11: 96 ▸ [Myanmar]

E. rufa Esben-Petersen, 1928. Treubia 10: 230 ▸ Indonesia: Sumatra

E. ruficauda Enderlein, 1910. Stettiner entomologische Zeitung 71: 365 ▸ Indonesia: Sulawesi

E. rufocincta Handschin, 1961. Nova Guinea Zoology 1: 270 ▸ Borneo

E. sauteri Navás, 1927. Entomologische Mitteilungen 16: 39 ▸ [Taiwan]

E. tagalensis Banks, 1914. Proceedings of the Entomological Society of Washington 15: 176 ▸ Philippines

E. triangularis Handschin, 1961. Nova Guinea Zoology 1: 265 ▸ Philippines

E. vespiformis Okamoto, 1911. Dobutsugaku Zasshi Tokyo 22: 541 ▸ [Taiwan]

Eumantispa Okamoto, 1910

E. araucariae Handschin, 1961. Nova Guinea Zoology 1: 295 ▸ New Guinea

E. bouchardi (Navás, 1909). Memorias de la Real Academia de Ciencias y Artes de Barcelona 7: 8 ▸ Indonesia: Sumatra

E. campioni (Navás, 1914). Memorias de la Real Academia de Ciencias y Artes de Barcelona 11: 97 ▸ Singapore [?India]

E. ferruginea (Stitz, 1913). Mitteilungen aus dem Zoologischen Museum in Berlin 7: 31 ▸ Indonesia: Sulawesi

E. fuscata Navás, 1914. Memorias de la Real Academia de Ciencias y Artes de Barcelona 11: 88 ▸ Indonesia: Sulawesi

**E. harmandi* (Navás, 1909). Memorias de la Real Academia de Ciencias y Artes de Barcelona 7: 10 ▸ [Taiwan; Japan]

E. lombokensis Handschin, 1961. Nova Guinea Zoology 1: 294 ▸ Indonesia: Lombok

E. moluccensis Handschin, 1961. Nova Guinea Zoology 1: 296 ▸ Indonesia: Moluccas

**E. taiwanensis* Kuwayama, 1925. Journal of the College of Agriculture, Hokkaido Imperial University 15: 258 ▸ [Taiwan]

Mantispa Illiger, 1798

**M. annulicornis* Gerstaecker, 1893. Mitteilungen aus dem Naturwissenschaftlichen Verein für Neu-Vorpommern und Rügen in Greifswald 25: 151 ▸ [Thailand]

**M. azihuna* Stitz, 1913. Mitteilungen aus dem Zoologischen Museum in Berlin 7: 7 ▸ [Taiwan]

M. basalis (Navás, 1929). Annali del Museo civico di storia naturale di Genova 53: 380 ▸ Indonesia: Ambon

**M. bicolor* Stitz, 1913. Mitteilungen aus dem Zoologischen Museum in Berlin 7: 8 ▸ [Vietnam]

M. celebensis Enderlein, 1910. Stettiner entomologische Zeitung 71: 348 ▸ Indonesia: Sulawesi

M. completa Banks, 1920. Bulletin of the Museum of Comparative Zoology at Harvard 64: 340 ▸ Philippines

**M. coomani* (Navás, 1930). Brotéria Série zoológica 26: 12 ▸ [Vietnam]

M. cordieri (Navás, 1933). Memorie della Pontificia Accademia delle Scienze, Nuovi Lincei (2) 17: 28 ▸ Indonesia: Java

M. decepta Banks, 1920. Bulletin of the Museum of Comparative Zoology at Harvard 64: 341 ▸ Philippines

M. enderleini Banks, 1914. Proceedings of the Entomological Society of Washington 15: 176 ▸ Philippines

M. javanica Westwood, 1852. The Transactions of the Entomological Society of London (2) 1: 267 ▸ Indonesia: Java

M. lineolata Westwood, 1853. The Transactions of the Entomological Society of London (2) 1: 267 ▸ Indonesia: Java, [Nepal]

M. luzonica Navás, 1909. Memorias de la Real Academia de Ciencias y Artes de Barcelona 7: 9 ▸ Indonesia: Sumatra; Philippines

M. melanocera Navás, 1913. Memorias de la Real Academia de Ciencias y Artes de Barcelona (10) 24: 26 ▸ Papua New Guinea

M. mimetica MacLachlan, 1900. The Entomologist's monthly magazine 36: 127 ▸ Borneo

M. moluccensis Banks, 1913. Transactions of the American Entomological Society 39: 208 ▸ Indonesia: Ambon

M. neptunica Navás, 1914. Memorias de la Real Academia de Ciencias y Artes de Barcelona 11: 212 ▸ Malaysia

M. newmanni Banks, 1920. Bulletin of the Museum of Comparative Zoology at Harvard 64: 340 ▸ Borneo

**M. nigra* Stitz, 1913. Mitteilungen aus dem Zoologischen Museum in Berlin 7: 3 ▸ [Taiwan]

M. pallescens Stitz, 1913. Mitteilungen aus dem Zoologischen Museum in Berlin 7: 22 ▸ Borneo

M. *pasteuri* Navás, 1909. Memorias de la Real Academia de Ciencias y Artes de Barcelona 7: 12 ‣ New Guinea

M. *plicicollis* Handschin, 1935. Revue suisse de zoologie et annales du Musée d'Histoire Naturelle de Genève 42: 704 ‣ Indonesia: Timor

M. *radiata* (Navás, 1914). Memorias de la Real Academia de Ciencias exactas, físicas y naturales de Madrid 12: 649 ‣ New Guinea

M. *simplex* Stitz, 1913. Mitteilungen aus dem Zoologischen Museum in Berlin 7: 23 ‣ Indonesia: Sulawesi

M. *taylori* Navás, 1927. Boletín de la Sociedad Ibérica de Ciencias Naturales 26: 57 ‣ Philippines

*M. *tonkinensis* Navás, 1930. Brotéria Série zoológica 26: 11 ‣ [Vietnam]

*M. *transversa* Stitz, 1913. Mitteilungen aus dem Zoologischen Museum in Berlin 7: 5 ‣ [Taiwan]

Mimetispa Handschin, 1961

M. *simulatrix* (MacLachlan, 1900). The Entomologist's monthly magazine 36: 127 ‣ Indonesia: Java; Borneo

Necyla Navás, 1913

N. *extrema* Navás, 1914. Revista de la Academia de Ciencias Exactas, Físico-Químicas y Naturales de Madrid 12: 481 ‣ New Guinea

N. *leopoldi* Navás, 1931. Mémoires du Musée Royal d'Histoire Naturelle de Belgique 4: 9 ‣ New Guinea

*N. *formosana* (Okamoto, 1911). Zoologischer Anzeiger 29: 297 ‣ [Taiwan]

*N. *orientalis* (Esben-Petersen, 1913). Entomologische Mitteilungen 2: 261 ‣ [Taiwan]

N. *trilineata* Navás, 1929. Memorie della Pontificia Accademia delle Scienze, Nuovi Lincei (2) 14: 12 ‣ Indonesia: Java

Orientispa Poivre, 1984

O. *shirozui* (Nakahara, 1961). Mushi 35: 63 ‣ Indonesia [Sri Lanka; Japan]

Stenomantispa Stitz, 1913

S. *ilsae* Stitz, 1913. Mitteilungen aus dem Zoologischen Museum in Berlin 7: 48 ‣ New Guinea

S. *reinhardi* Stitz, 1913. Mitteilungen aus dem Zoologischen Museum in Berlin 7: 24 ‣ New Guinea

Tuberonotha Handschin, 1961

T. *regia* (Navás, 1930). Memorie della Pontificia Accademia delle Scienze, Nuovi

Lincei (2) 14: 425 ▸ Indonesia: Sumatra; Malaysia; Borneo [India]
T. strenua (Gerstaecker, 1893). Mitteilungen aus dem Naturwissenschaftlichen Verein für Neu-Vorpommern und Rügen in Greifswald 25: 150
▸ Indonesia: Irian Jaya; Philippines [Japan]

Xaviera Lambkin, 1986
X. manca (Gerstaecker, 1885). Mitteilungen aus dem Naturwissenschaftlichen Verein für Neu-Vorpommern und Rügen in Greifswald 16: 39
▸ Indonesia; Papua New Guinea [Australia]

Berothidae

BEROTHINAE

Berotha Walker, 1860
B. piepersi van der Weele, 1904. Notes from the Leyden Museum 24: 214
▸ Indonesia: Java; Singapore

Frawalkeria Navás, 1929
F. borneensis (Navás, 1913). Memorias de la Real Academia de Ciencias y Artes de Barcelona 10: 66 ▸ Borneo
**F. puncticollis* (Navás, 1911). Revue russe d'entomologie 11: 112 ▸ [Taiwan]

Isoscelipteron A. Costa, 1863
I. borneole Aspöck & Aspöck, 1991. Zeitschrift der Arbeitsgemeinschaft Österreichischer Entomologen 43: 71 ▸ E. Malaysia: Sarawak
**I. tonkinense* (Krüger, 1922). Stettiner entomologische Zeitung 83: 65
▸ [Vietnam]
**I. formosensis* (Krüger, 1922). Stettiner entomologische Zeitung 83: 66 ▸ [Taiwan]
I.leveri (Kimmins, 1951). The Annals and Magazine of Natural History, including Zoology, Botany and Geology 12 (4): 783 ▸ Solomon Islands
I. nicobaricum (Navás, 1912). Brotéria Série zoológica 10: 108
▸ Malaysia; Indonesia; Krakatau Islands [Nicobar Islands]
I. philippinicola Aspöck & Aspöck, 1991. Zeitschrift der Arbeitsgemeinschaft Österreichischer Entomologen 43: 69 ▸ Philippines

Dilaridae

Berothella Banks, 1934

B. *phantoma* Banks, 1934. Journal of the Federated Malay States Museums 17: 568 ▸ Malaysia

Dilar Rambur, [1838]

D. *marmoratus* (Banks, 1931). Journal of the Federated Malay States Museums 16: 385 ▸ Thailand

D. *grandis* (Banks, 1931). Journal of the Federated Malay States Museums 16: 413 ▸ Borneo

*D. *formosanus* (Okamoto & Kuwayama, 1920). Dobutsugaku Zasshi Tokyo 32: 341 ▸ [Taiwan]

*D. *taiwanensis* Banks, 1937. The Philippine journal of science 62: 276 ▸ [Taiwan]

*D. *distinctus* Nakahara, 1955. Kontyû 23: 139 ▸ [Taiwan]

*D. *kanoi* (Nakahara, 1955). Kontyû 23: 6 ▸ [Taiwan]

*D. *pallidus* Nakahara, 1955. Kontyû 23: 140 ▸ [Taiwan]

D. *macleodi* Oswald & Schiff, 2001. Proceedings of the Entomological Society of Washington 103: 74-80 ▸ Malaysia: Sarawak

Hemerobiidae

Drepanacra Tillyard, 1916

*D. *plaga* Banks, 1939. Bulletin of the Museum of Comparative Zoology at Harvard 85: 467 ▸ [Taiwan]

Hemerobius Linnaeus, 1758

H. *baguiensis* Navás, 1923. Memorie della Pontificia Accademia delle Scienze, Nuovi Lincei (2) 6: 14 ▸ Philippines

*H. *cercodes* Navás, 1917. Insecta : revue illustrée d'entomologie; publication mensuelle de la Station Entomologique de la Faculté des Sciences de Rennes 7: 14 ▸ [Vietnam]

H. *claggi* Banks, 1937. The Philippine journal of science 63: 141 ▸ Philippines

*H. *hyalinus* Nakahara, 1966. Kontyû 34: 210 ▸ [Taiwan]

H. *incursus* Banks, 1930. Journal of the Federated Malay States Museums 16: 384. ▸ Malaysia

H. *inversus* Navás, 1927. Memorie della Pontificia Accademia delle Scienze, Nuovi Lincei (2) 10: 21 ▸ Philippines

*H. *japonicus* Nakahara, 1915. Annotationes Zoologicae Japonenses 9: 25 ▸ [Taiwan; Japan]

H. kutsimensis New, 1989 (1988). Invertebrate Taxonomy 2: 614
▸ Indonesia: Irian Jaya
H. morobensis New, 1989 (1988). Invertebrate Taxonomy 2: 613
▸ Papua New Guinea
H. ricarti Navás, 1925. Memorie della Pontificia Accademia delle Scienze, Nuovi Lincei (92) 6: 13 ▸ Philippines
H. rizali Banks, 1920. Bulletin of the Museum of Comparative Zoology at Harvard 64: 332 ▸ Philippines
H. spinigerus Banks, 1937. The Philippine journal of science 62: 279 ▸ Philippines
H. sumatranus Navás, 1926. Memorie della Pontificia Accademia delle Scienze, Nuovi Lincei (2) 9: 116 ▸ Indonesia: Sumatra
H. tagalicus Banks, 1920. Bulletin of the Museum of Comparative Zoology at Harvard 64: 332 ▸ Philippines
H. vagans Banks, 1937. The Philippine journal of science 63: 141 ▸ Philippines
H. sp. New, 1989 (1988). Invertebrate Taxonomy 2: 614 ▸ Papua New Guinea

Megalomus Rambur, 1842
**M. formosanus* Banks, 1937. The Philippine journal of science 62: 279 ▸ [Taiwan]

Micromus Rambur, 1842
**M. alternans* Navás, 1923. Arxius de la secció de ciències. Institud d'estudis Catalans. Barcelona 7: 194 ▸ [Vietnam]
M. angustior van der Weele, 1910. Notes from the Leyden Mus. 32: 202
▸ Indonesia: Java; Philippines
**M. bernardi* Navás, 1911. Verhandlungen des VIII. internationalen Zoologen-Congresses zu Graz 15.-20. August 1910: 750 ▸ [Vietnam]
M. brandti New, 1989 (1988). Invertebrate Taxonomy 2: 616 ▸ Papua New Guinea
**M. duporti* Lestage, 1929. Bulletin et annales de la Société Entomologique de Belgique 68: 265 ▸ [Vietnam]
M. gratus Banks, 1937. The Philippine journal of science 63: 138 ▸ Philippines
M. igorotus Banks, 1920. Bulletin of the Museum of Comparative Zoology at Harvard 64: 335 ▸ Philippines
M. jacobsoni Esben-Petersen, 1926. Entomologische Mitteilungen 15: 406
▸ Indonesia: Sumatra
**M. kanoi* (Nakahara, 1955). Kontyû 23: 8 ▸ [Taiwan]
M. lorianus (Navás, 1929). Annali del Museo civico di storia naturale di Genova 53: 374 ▸ Papua New Guinea
M. maculatipes (Nakahara, 1915). Annotationes Zoologicae Japonenses 9: 39
▸ Philippines [Japan]

M. *morosus* Gerstaecker, 1894. Mitteilungen aus dem Naturwissenschaftlichen Verein für Neu-Vorpommern und Rügen in Greifswald 25: 170 ▸ Indonesia: Java

M. *multipunctatus* Matsumura, 1907. Systematic Entomology, vol 1 ('Konchu Bunruigaku' 1) Tokyo: 171 ▸ Indonesia: Java; Philippines [Japan; China]

M. *nigrifrons* Banks, 1937. The Philippine journal of science 63: 139 ▸ Philippines

M. *philippinus* Navás, 1926 (=M. *timidus*). Memorie della Pontificia Accademia delle Scienze, Nuovi Lincei 2 (9): 117 ▸ Philippines

M. *pictipes* Banks, 1920 (= M. *maculatipes*). Bulletin of the Museum of Comparative Zoology at Harvard 64: 334 ▸ Philippines

M. *placidulus* Banks, 1937. The Philippine journal of science 63: 139 ▸ Philippines

M. *timidus* Hagen, 1853. Bericht über die Verhandlungen die K. Preussischen Akademie der Wissenschaften zu Berlin (1853): 481 ▸ Indonesia: Java; Papua New Guinea; Philippines [widespread, from Africa to Pacific islands, and with many synonyms]

Neuronema MacLachlan, 1869

*N. *navasi* Kimmins, 1943. The Annals and Magazine of Natural History, including Zoology, Botany and Geology 11 (10): 49 ▸ [Taiwan]

Noius Navás, 1929

NOTE

One species has been reported simply as '*Noius* sp.':

Noius sp. Oswald, 1993. Journal of the New York Entomological Society 101: 249 ▸ New Guinea

Notobiella Banks, 1909

N. *dentata* Monserrat, 1990. Proceedings of the 3rd International Symposium on Neuropterology : 68 ▸ Indonesia

N. *gressitti* New, 1989 (1988). Invertebrate Taxonomy 2: 608 ▸ Indonesia: Irian Jaya; Papua New Guinea

N. *luisae* Monserrat, 1990. Proceedings of the 3rd International Symposium on Neuropterology: 68 ▸ Indonesia

N. *multifurcata* Tillyard, 1916. Proceedings of the Linnean Society of New South Wales 1: 310 ▸ Papua New Guinea [Australia; New Caledonia]

N. *ocellata* New, 1989 (1988). Invertebrate Taxonomy 2: 607 ▸ Papua New Guinea

*N. *ochracea* Nakahara, 1966. Kontyû 34: 195 ▸ [Taiwan]

N. *sedlaceki* New, 1989 (1988). Invertebrate Taxonomy 2: 608 ▸ Papua New Guinea

N. *tumida* (Navás, 1925). Memorias de la Real Academia de Ciencias y Artes de Barcelona 19: 14 ▸ Indonesia: Java

N. *unita* Banks, 1909. Proceedings of the Entomological Society of Washington 11: 80 ▸ Borneo

N. *valida* Banks, 1920. Bulletin of the Museum of Comparative Zoology 64: 334 ▸ Borneo

N. *viridis* Tillyard, 1916. Proceedings of the Linnean Society of New South Wales 41: 309 ▸ Papua New Guinea; [Australia]

Psectra Hagen, 1866

P. *affinis* (Banks, 1913). Transactions of the American Entomological Society 39: 219 ▸ Philippines

*P. *decorata* (Nakahara, 1966). Kontyû 34: 196 ▸ [Taiwan]

P. *franzeni* (Kimmins, 1940). The Annals and Magazine of Natural History, including Zoology, Botany and Geology 11 (6): 231 ▸ Indonesia: Irian Jaya; Papua New Guinea

P. *latilobata* New, 1989 (1988). Invertebrate Taxonomy 2: 611 ▸ Papua (Irian Jaya)

P. *minima* (Banks, 1920). Bulletin of the Museum of Comparative Zoology at Harvard 64: 333 ▸ Borneo

P. *oriomoense* New, 1989 (1988). Invertebrate Taxonomy 2: 611 ▸ Papua New Guinea

P. *siamica* Nakahara & Kuwayama, 1961. Nature and life in Southeast Asia 1: 261 ▸ ▸ Thailand

P. *wilhelmense* New, 1989 (1988). Invertebrate Taxonomy 2: 610 ▸ Papua New Guinea

Wesmaelius Krüger, 1922

W. *davidicus* (Navás, 1910). Russkoe entomologiceskoe obozrenie 10: 192 ▸ Philippines

Zachobiella Banks, 1920

Z. *jacobsoni* Esben-Petersen, 1926. Entomologische Mitteilungen 15: 407 ▸ Indonesia: Sumatra

Z. *marmorata* Navás, 1926. Memorie della Pontificia Accademia delle Scienze, Nuovi Lincei 2 (9): 118 ▸ Indonesia: Sumatra

Z. *pallida* Banks, 1939. Bulletin of the Museum of Comparative Zoology at Harvard 85: 466 ▸ Papua New Guinea [Australia]

Z. *punctata* Banks, 1920. Bulletin of the Museum of Comparative Zoology at Harvard 64: 335 ▸ Philippines

*Z. *striata* Nakahara, 1966. Kontyû 34: 198 ▸ [Taiwan]

Chrysopidae

APOCHRYSINAE

Joguina Navás, 1912
J. borneensis Kimmins, 1952. The Annals and Magazine of Natural History, including Zoology, Botany and Geology 12 (5): 942 ▸ Borneo
J. malayana Banks, 1931. Journal of the Federated Malay States Museums 16: 383. ▸ Malaysia

Nacaura Navás, 1913
**N. matsumurae* (Okamoto, 1912). Transactions - Sapporo Natural History Society 4: 13 ▸ [Taiwan; Japan]

Nobilinus Navás, 1913
N. *albardae* (MacLachlan, 1875). Tijdschrift voor Entomologie 18: 3. ▸ Indonesia; Papua New Guinea; Philippines: Palawan
N. *bellula* (Banks, 1914). Proceedings of the Entomological Society of Washington 15: 175 ▸ Philippines
N. *coccinea* (Brauer, 1864). Novara-Expedition. Zoologischer Theil. Band 1. Neuropteren: 30 ▸ Indonesia: Ambon

Synthochrysa Needham, 1909
**S. cognata* Kimmins, 1953. The Annals and Magazine of Natural History, including Zoology, Botany and Geology 12 (6): 246 ▸ [New Hebrides]
S. evanida (Gerstaecker, 1893). Mitteilungen aus dem Naturwissenschaftlichen Verein für Neu-Vorpommern und Rügen in Greifswald 25: 154 ▸ Indonesia: Java
**S. montrouzieri* (Girard, 1862). Annales de la Societé entomologique de France 4 (2): 609 ▸ [Loyalty Islands]
S. salomonis (Kimmins, 1951). The Annals and Magazine of Natural History, including Zoology, Botany and Geology 12 (4): 780 ▸ Solomon Islands

CHRYSOPINAE

Ankylopterygini

Ankylopteryx Brauer, 1864
A. braueri Banks,1937. The Philippine journal of science 63: 142 ▸ Philippines
A. doleschalii Brauer, 1864. Verhandlungen der Zoologisch-Botanischen Gesellschaft in Wien 14: 901 ▸ Indonesia: Ambon

*A. *gracilis* Nakahara, 1955. Kontyû 23: 143 ▸ [Taiwan]

A. nervosa Navás, 1913. Annales de la Société Scientifique de Bruxelles 37: 293 ▸ Malaysia: Sarawak

A. nonelli Navás, 1913. Annales de la Société Scientifique de Bruxelles 37: 294 ▸ Singapore

A. obliqua Banks, 1924. Bulletin of the Museum of Comparative Zoology at Harvard 65: 434 ▸ Philippines

A. octopunctata (Fabricius, 1793). Entomologia Systematica emandata et aucta 2: 85 ▸ Indonesia; Malaysia

A. perpallida Banks, 1924. Bulletin of the Museum of Comparative Zoology at Harvard 65: 434 ▸ Philippines

A. polygramma Gerstaecker, 1893. Mitteilungen aus dem Naturwissenschaftlichen Verein für Neu-Vorpommern und Rügen in Greifswald 25: 161 ▸ Indonesia: Java

*A. (*Sencera*) *feae* Navás, 1929. Annali del Museo civico di storia naturale di Genova 53: 371 ▸ [Myanmar]

*A. (*S.*) *exquisita* Nakahara, 1955. Kontyû 23: 143 ▸ [Taiwan]

A. (S.) scioneura Navás, 1924. Revista de la Academia de ciencias exactas, físicas químicas y naturales de Zaragoza 9: 27 ▸ Indonesia; Malaysia; Papua New Guinea; Philippines

Retipenna Brooks, 1986

*R. *burmana* Brooks, 1986. Neuroptera International 4: 41 ▸ [Myanmar]

*R. *hasegawai* (Nakahara, 1955). Kontyû 23: 145 ▸ [Taiwan]

R. notata (Navás, 1910). Brotéria Série zoológica 9: 40 ▸ Malaysia; Philippines [India; Myanmar]

Semachrysa Brooks, 1983

S. claggi (Banks, 1937). The Philippine journal of science 63: 143 ▸ Philippines

S. cruciata (Esben-Petersen, 1928). Treubia 10: 228 ▸ Indonesia: Sumatra

S. dammermanni (Esben-Petersen, 1929). Treubia, Supplement 7: 103 ▸ New Guinea

S. decorata (Esben-Petersen, 1913). Entomologische Mitteilungen 2: 260 ▸ Malaysia; Philippines [Taiwan; Japan; China]

*S. *matsumurae* (Okamoto, 1914). Journal of the College of Agriculture, Tohoku Imperial University 6: 68 ▸ [Taiwan; Japan; China]

S. minuta (Brooks, 1983). Bulletin of the British Museum (Natural History) / Entomology 47: 11 ▸ Indonesia; Malaysia [Myanmar]

S. nigribasis (Banks, 1920). Bulletin of the Museum of Comparative Zoology at Harvard 64: 337 ▸ Malaysia

S. papuensis Brooks, 1983. Bulletin of the British Museum (Natural History) / Entomology 47: 12 ▸ New Guinea
S. wallacei Brooks, 1983. Bulletin of the British Museum (Natural History) / Entomology 47: 24 ▸ New Guinea

Belonopterygini

Evanochrysa Brooks & Barnard, 1990
E. evanescens (MacLachlan, 1869). The Entomologist's monthly magazine 6: 25 ▸ Indonesia; Malaysia

Italochrysa Principi, 1946
I. aequalis (Walker, 1853). Catalogue of the specimens of Neuropterous insects in the collection of the British Museum II: 266 ▸ Indonesia; Papua New Guinea
I. ferruginea (MacLachlan, 1869). The Entomologist's monthly magazine 6: 26 ▸ Malaysia: Sabah; Indonesia
I. froggatti (Esben-Petersen, 1914). Proceedings of the Linnean Society of New South Wales 39: 640 ▸ Papua New Guinea [Australia]
I. gillavryi (Navás, 1924). Brotéria Série zoológica 21: 128 ▸ Indonesia: Java
I. ignobilis (Navás, 1913). Annales de la Société Scientifique de Bruxelles 37: 322 ▸ Singapore
I. jubilaris (Navás, 1925). Revista de la Academia de ciencias exactas, físicas químicas y naturales de Zaragoza (1924) 9: 29 ▸ Papua New Guinea
**I. levasseuri* (Navás, 1921). Revista de la Academia de ciencias exactas, físicas químicas y naturales de Zaragoza 6: 74 ▸ [Vietnam]
I. ludekingi (van der Weele, 1909). Notes from the Leyden Museum 31: 77 ▸ Indonesia
**I. modesta* Nakahara, 1955. Kontyû 23: 146 ▸ [Taiwan]
I. neobritannica (Navás, 1913). Memorias de la Real Academia de Ciencias y Artes de Barcelona 10: 24 ▸ Papua New Guinea: New Britain
I. pallicornis (Banks, 1920). Bulletin of the Museum of Comparative Zoology at Harvard 64: 336 ▸ Singapore
I. polychroa (Gerstaecker, 1893). Mitteilungen aus dem Naturwissenschaftlichen Verein für Neu-Vorpommern und Rügen in Greifswald 25: 163 ▸ Indonesia: Java
I. simplex (Banks, 1920). Bulletin of the Museum of Comparative Zoology at Harvard 64: 336 ▸ Borneo
**I. subcostalis* (Navás, 1917). Insecta : revue illustrée d'entomologie; publication mensuelle de la Station Entomologique de la Faculté des Sciences de Rennes 7: 17 ▸ [Vietnam]

I. sumatrana (Albarda, 1881). Midden-Sumatra 4: 15 ▸ Indonesia: Sumatra
I. tibialis (Navás, 1913). Annales de la Société Scientifique de Bruxelles 37: 321 ▸ Indonesia: Sumatra
**I. uchidae* (Kuwayama, 1927). Insecta Matsumurana 1: 120 ▸ [Taiwan]

Nodochrysa Banks, 1938
N. necrota (Banks, 1938). Journal of the Federated Malay States Museums 18: 226 ▸ Malaysia; Singapore, Brunei

Stigmachrysa Navás, 1925
S. elegans Esben-Petersen, 1933. Videnskabelige meddelser fra Dansk naturhistorisk forening i København 94: 120 ▸ Singapore
S. kervillei Navás, 1925. Annales de la Société Scientifique de Bruxelles 44: 569 ▸ Indonesia: Java

Chrysopini

Apertochrysa Tjeder, 1966
A. physophlebia (Navás, 1913). Annales de la Société Scientifique de Bruxelles 38: 99 ▸ Malaysia; Indonesia

Austrochrysa Esben-Petersen, 1928
A. abnormis (Albarda, 1881). Midden-Sumatra 4: 16 ▸ Indonesia: Sumatra, Java
A. apoana (Banks, 1937). The Philippine journal of science 63: 146 ▸ Philippines
A. loriana (Navás, 1929). Annali del Museo civico di storia naturale di Genova 53: 365 ▸ New Guinea

Borniochrysa Brooks & Barnard, 1990
B. appendiculata (Esben-Petersen, 1926). Entomologische Mitteilungen 15: 26 ▸ Indonesia: Sumatra
B. luzonica (Banks, 1939). The Philippine journal of science 69: 138 ▸ Philippines
B. solomonis (Banks, 1941). American Museum Novitates 1143: 5 ▸ Papua New Guinea: Solomon Islands
B. winkleri (Navás, 1928). Boletín de la Sociedad Entomológica de España 11: 124 ▸ ▸ Borneo

Chrysocerca van der Weele, 1909
**C. formosana* (Okamoto, 1914). The Journal of the Sapporo Agricultural College 6: 55 ▸ [Taiwan]

C. jacobsoni (van der Weele, 1909). Notes from the Leyden Museum 31: 65 ▸ Indonesia: Java

C. timorina (Handschin, 1935). Revue suisse de zoologie et annales du Musée d'Histoire Naturelle de Genève 42: 694 ▸ Indonesia: Timor

Chrysoperla Steinmann, 1964

C. congrua (Walker, 1853). Catalogue of the specimens of Neuropterous insects in the collection of the British Museum II: 238 ▸ W. Malaysia; Thailand; Indonesia [Africa; Sri Lanka; Australia; much of western Pacific]

C. furcifera (Okamoto, 1914). Journal of the College of Agriculture, Tohoku Imperial University 6: 61 ▸ Philippines [Taiwan; Japan; China]

C. nipponensis (Okamoto, 1914). Journal of the College of Agriculture, Tohoku Imperial University 6: 65 ▸ Philippines [Japan; China; Korea]

C. oblita (Hölzel, 1973). Khumbu Himal 4: 342 ▸ Thailand [Nepal; India]

C. siamensis Brooks, 1994. Bulletin of the British Museum (Natural History) / Entomology 63: 159 ▸ Thailand

Chrysopa Leach, 1815

NOTE

The following species were placed as 'incertae sedis' by Brooks & Barnard (1990), and thus reflect the status of *Chrysopa* as a holding genus. Names in parentheses are tentative generic allocations suggested by Brooks & Barnard (1990).

Chrysopa s.l.

C. adnexa Navás, 1929. Annali del Museo civico di storia naturale di Genova 53: 357 ▸ Indonesia: Java; New Guinea

C. adnixa Esben-Petersen, 1913. (?*Mallada*). Notes from the Leyden Museum 35: 232 ▸ Indonesia: Java, Bali

C. adonis Banks, 1937. The Philippine journal of science 62: 282 ▸ [Taiwan]

C. atrioris Banks, 1920. Bulletin of the Museum of Comparative Zoology at Harvard 64: 237 ▸ Singapore; Borneo; Philippines

C. aurea Kimmins, 1951. The Annals and Magazine of Natural History, including Zoology, Botany and Geology 12 (4): 782 ▸ Papua New Guinea: Solomon Islands

C. azygota Banks, 1915. Proceedings of the Academy of Natural Sciences of Philadelphia 66: 628 ▸ Philippines

C. beccarii Navás, 1929. Annali del Museo civico di storia naturale di Genova 53: 358 ▸ New Guinea

C. *buehleri* Handschin, 1936. Verhandlungen der Naturforschenden Gesellschaft in Basel 47: 133 ▸ Indonesia: Rote

C. *caprae* Navás, 1929. Annali del Museo civico di storia naturale di Genova 53: 366 ▸ New Guinea

C. *dahli* Navás, 1925. Revista de la Academia de ciencias exactas, físicas químicas y naturales de Zaragoza 9: 21 ▸ New Guinea: New Britain (?*Plesiochrysa*)

C. *derota* Banks, 1937. The Philippine journal of science 63: 145 ▸ Philippines

*C. *eudora* Banks, 1937. The Philippine journal of science 62: 283 ▸ Philippines

C. *excelsior* Banks, 1937. The Philippine journal of science 63: 144 ▸ Philippines

*C. *feana* Navás, 1929. Annali del Museo civico di storia naturale di Genova 53: 367 ▸ [Taiwan]

C. *frequens* Esben-Petersen, 1913. Notes from the Leyden Museum 35: 232 ▸ Indonesia: Java, Timor, Rote

C. *gestroi* Navás, 1929. Annali del Museo civico di storia naturale di Genova 53: 360 ▸ New Guinea

C. *isolata* Banks, 1914. Proceedings of the Entomological Society of Washington 15: 174 ▸ Philippines

C. *javanica* Esben-Petersen, 1913 (?*Mallada*). Notes from the Leyden Museum 35: 230 ▸ Indonesia: Java

C. *julia* Navás, 1927. Entomologische Mitteilungen 16: 37 ▸ Indonesia: Java

C. *malayana* Esben-Petersen, 1926. Entomologische Mitteilungen 15: 28 ▸ Indonesia: Sumatra

*C. *marcida* Banks, 1937. The Philippine journal of science 62: 284 ▸ [Taiwan]

C. *mesonotalis* Esben-Petersen, 1926 (?*Chrysocerca*). Entomologische Mitteilungen 15: 27 ▸ Indonesia: Sumatra

C. *meyeri* Handschin, 1936. Verhandlungen der Naturforschenden Gesellschaft in Basel 47: 132 ▸ Indonesia: Timor, Rote

C. *mindanensis* Banks, 1937. The Philippine journal of science 63: 146 ▸ Philippines

C. *notulata* Banks, 1937. The Philippine journal of science 63: 145 ▸ Philippines

C. *ochracea* Albarda, 1881. Midden-Sumatra 4: 15 ▸ Indonesia: Sumatra

*C. *parallela* Navás, 1931. Revista de la Academia de Ciencias Exactas, Físico-Químicas y Naturales de Madrid 26: 82 ▸ [Myanmar]

C. *pigmentata* Handschin, 1935. Revue suisse de zoologie et annales du Musée d'Histoire Naturelle de Genève 42: 698 ▸ Indonesia: Flores, Timor

C. *sumatrensis* (Navás, 1929) (? *Austrochrysa*). Memorie della Pontificia Accademia delle Scienze, Nuovi Lincei (92) 12: 35 ▸ Indonesia: Sumatra

C. *thieli* Navás, 1929 (?*Plesiochrysa*). Boletín de la Sociedad Entomológica de España 12: 73 ▸ Papua New Guinea

C. *valdezi* Banks, 1924 (?*Signochrysa*). Bulletin of the Museum of Comparative

Zoology at Harvard 65: 433 ▸ Borneo

C. virgata Handschin, 1935. Revue suisse de zoologie et annales du Musée d'Histoire Naturelle de Genève 42: 697 ▸ Indonesia: Java

Chrysopidea Navás, 1910

C. (Chrysotropia) obliquata (Banks, 1931). Journal of the Federated Malay States Museums 16: 64 ▸ Philippines

Glenochrysa Esben-Petersen, 1920

G. splendida (van der Weele, 1909). Notes from the Leyden Museum 31: 72 ▸ Indonesia; Malaysia

Kostka Navás, [1913]

K. nacaratus Navás, 1913. Annales de la Société Scientifique de Bruxelles 37: 319 ▸ Indonesia: Sumatra

Mallada Navás, [1925]

NOTE

Species of *Mallada* and *Dichochrysa* are listed uncritically as *Mallada* here.

M. alticola (Banks, 1931). Journal of the Federated Malay States Museums 16: 414 ▸ Borneo

M. anpingensis (Esben-Petersen, 1913). Entomologische Mitteilungen 2: 259 ▸ [Taiwan]

M. basalis (Walker, 1853). Catalogue of the specimens of Neuropterous insects in the collection of the British Museum II: 239 ▸ Indonesia; Philippines; Papua New Guinea [Australia; widespread in western Pacific]

M. crassoneura (van der Weele, 1909). Notes from the Leyden Museum 31: 71 ▸ Indonesia; Malaysia

M. eurycista (Navás, 1914). Annales de la Société Scientifique de Bruxelles 38: 101 ▸ ▸ Singapore

M. flaveola (Schneider, 1851). Symbolae ad monographiam generis Chrysopae, Leach. Vratislavae: 75 ▸ Indonesia: Java

M. krakatauensis Tsukaguchi, 1988. Kontyû 56: 482 ▸ Indonesia: Krakatau Islands

M. macquilingi (Banks, 1937). The Philippine journal of science 63: 149 ▸ Philippines

M. morota (Banks, 1915). Proceedings of the Academy of Natural Sciences of Philadelphia 66: 629 ▸ Philippines

M. nea (Navás, 1912). Brotéria Série zoológica 10: 106 ▸ New Guinea

M. neglecta (Banks, 1931). Journal of the Federated Malay States Museums 16: 382 ▸ Malaysia

M. notalis (Navás, 1914). Annales de la Société Scientifique de Bruxelles 37: 101 ▸ Indonesia: Sulawesi

M. signatus (Schneider, 1851). Symbolae ad monographiam generis Chrysopae, Leach. Vratislavae: 109 ▸ Indonesia; Papua New Guinea [Australia]

M. sumatrensis (Esben-Petersen, 1926). Entomologische Mitteilungen 15: 25 ▸ Indonesia: Sumatra

Plesiochrysa Adams, 1982

P. ramburi (Schneider, 1851). Symbolae ad monographiam generis Chrysopae, Leach. Vratislavae: 107 ▸ Indonesia; Malaysia; Papua New Guinea [Australia; widespread in western Pacific]

P. ruficeps (MacLachlan, 1875). Tijdschrift voor Entomologie 18: 2 ▸ Indonesia: Sulawesi, Flores; Papua New Guinea

Nymphidae

Myiodactylus Brauer, 1866

M. chrysopoides Navás, 1922. Annales de la Société Scientifique de Bruxelles 41: 131 ▸ Philippines

M. lyriformis New, 1987. Invertebrate Taxonomy 1: 813 ▸ Papua New Guinea

M. nebulosus Maclachlan, 1877. The Entomologist's monthly magazine 14: 85 ▸ New Guinea [?Australia]

M. paranebulosus New, 1987. Invertebrate Taxonomy 1: 810 ▸ Indonesia: Irian Jaya

M. sorongensis New, 1987. Invertebrate Taxonomy I: 811 ▸ Indonesia: Irian Jaya

Osmylops Banks, 1913

O. hirsutus New, 1985. Neuroptera International 3: 187 ▸ Papua New Guinea

O. monofoveatus Oswald, 1998. Journal of Neuropterology 1: 102 ▸ Papua New Guinea

O. nesos Oswald, 1998. Journal of Neuropterology 1: 103 ▸ Papua New Guinea

Umbranymphes New, 1987

U. spinosus New, 1987. Invertebrate Taxonomy 1: 814 ▸ Papua (Irian Jaya)

Ascalaphidae

HAPLOGLENIINAE

Idricerus MacLachlan, [1871]
I. obscurus (Westwood, 1848). The Cabinet of Oriental Entomology, plate 34
▸ 'E. Indies'

Protidricerus van der Weele, [1909]
P. exilis (MacLachlan, 1894). The Annals and Magazine of Natural History, including Zoology, Botany and Geology 6 (13): 424 ▸ Borneo [China]
P. irene van der Weele, 1909. Collections Zoologiques du baron Edm. de Selys-Longchamps. Catalogue systématique et descriptif 8: 315 ▸ Borneo
P. philippinensis Esben-Petersen, 1927. Proceedings of the Zoological Society of London 1927: 550 ▸ Philippines

ASCALAPHINAE

Acheron Lefèbvre, 1842
A. trux (Walker, 1853). Catalogue of the specimens of Neuropterous insects in the collection of the British Museum II: 432
▸ W. Malaysia [Taiwan; China; India]

Ascalohybris Sziraki, 1999
A. angulata (Westwood, 1848). The Cabinet of Oriental Entomology
▸ W. Malaysia [India, Myanmar, Bangladesh]
A. borneensis (van der Weele, 1904). Notes from the Leyden Museum 24: 209
▸ Borneo
A. ferruginea (Navás, 1911). Mitteilungen der Münchener Entomologischen Gesellschaft 1911: 27 ▸ Indonesia: Sumatra
A. flavicans (van der Weele, 1904). Notes from the Leyden Museum 24: 208
▸ Indonesia: Sumatra
A. javana (Burmeister, 1839). Handbuch der Entomologie 2: 1001
▸ Indonesia: Java
A. malaccensis (van der Weele, 1909). Collections Zoologiques du baron Edm. de Sélys-Longchamps. Catalogue systématique et descriptif 8: 237
▸ Malaysia, Singapore
A. obscura (van der Weele, 1909). Collections Zoologiques du baron Edm. de Sélys-Longchamps. Catalogue systématique et descriptif 8: 233 ▸ Indonesia
**A. oberthuri* (Navás, 1923). Arxius de la secció de ciències. Institud d'estudis Catalans. Barcelona 7: 181 ▸ [Vietnam]

A. rufescens (Navás, 1911). Mitteilungen der Münchener Entomologischen Gesellschaft 2: 6 ▸ ?Indonesia: Sumatra (later inference)

**A. subjacens* (Walker, 1853). Catalogue of the specimens of Neuropterous insects in the collection of the British Museum II: 431
▸ [Vietnam; Taiwan; Japan; China]

**A. vitalisi* (Navás, 1914). Insecta : revue illustrée d'entomologie; publication mensuelle de la Station Entomologique de la Faculté des Sciences de Rennes 4: 133 ▸ [Cambodia]

Glyptobasis MacLachlan, [1871]

**G. brunnea* Esben-Petersen, 1913. Entomologische Mitteilungen 2: 224
▸ [Taiwan]

**G. fraseri* Sziráki, 1999, nom. nov. for *G. brunnea* Fraser, 1922. Records of the Indian Museum 24: 517 ▸ [Myanmar]

**G. spinicornis* van der Weele, 1909. Notes from the Leyden Museum 30: 245
▸ [Myanmar]

Haploglenius Burmeister, 1839

?*H. neoguineensis* Navás, 1913. Revue russe d'entomologie 13: 424
▸ Papua New Guinea

Helicomitus MacLachlan, [1871]

H. dicax (Walker, 1853). Catalogue of the specimens of Neuropterous insects in the collection of the British Museum II: 423
▸ Malaysia; Indonesia; Philippines [Taiwan]

Nousera Navás, 1923

**N. gibba* Navás, 1923. Memorie della Pontificia Accademia delle Scienze, Nuovi Lincei (2) 6: 5 ▸ [Indochina]

Ogcogaster Westwood, [1847]

**O. kirbyi* van der Weele, 1909. Collections Zoologiques du baron Edm. de Sélys-Longchamps. Catalogue systématique et descriptif 8: 254 ▸ [Cambodia]

Protacheron van der Weele, [1909]

P. philippinensis van der Weele, 1904. Notes from the Leyden Museum 24: 206
▸ Indonesia: Java, Sulawesi; Philippines

Suhpalacsa Lefèbvre, 1842

S. dietrichiae (Brauer, 1869). Verhandlungen der Zoologisch-Botanischen Gesellschaft in Wien 18: 15 ▸ Papua New Guinea [Australia]

S. donckieri (Navás, 1913). Arxius de la secció de ciències. Institud d'estudis Catalans. Barcelona 1: 88 ▸ Indonesia: Java

**S. formosana* (Okamoto, 1909). Wiener Entomologische Zeitung 29: 65 ▸ [Taiwan]

S. ledrana (Navás, 1913). Revue russe d'entomologie 13: 425 ▸ Indonesia: Java

S. lugubris (Gerstaecker, 1893). Mitteilungen aus dem Naturwissenschaftlichen Verein für Neu-Vorpommern und Rügen in Greifswald 25: 104 ▸ Indonesia: Java

S. princeps (Gerstaecker, 1893). Mitteilungen aus dem Naturwissenschaftlichen Verein für Neu-Vorpommern und Rügen in Greifswald 25: 103 ▸ Indonesia: Java

S. reducta Banks, 1931. Psyche 38: 63 ▸ Philippines

S. subtrahens (Walker, 1853). Catalogue of the specimens of Neuropterous insects in the collection of the British Museum II: 430 ▸ Papua New Guinea [Australia]

S. sumbawana van der Weele, 1909. Collections Zoologiques du baron Edm. de Sélys-Longchamps. Catalogue systématique et descriptif 8: 166 ▸ Indonesia: Sumba

**S. umbrosa* (Esben-Petersen, 1913). Entomologische Mitteilungen 2: 226 ▸ [Taiwan]

Suphalomitus van der Weele, [1909]

**S. formosanus* Esben-Petersen, 1913. Entomologische Mitteilungen 2: 225 ▸ [Taiwan]

S. harmandi van der Weele, 1909. Collections Zoologiques du baron Edm. de Sélys-Longchamps. Catalogue systématique et descriptif 8: 184 ▸ Thailand

S. jentinki van der Weele, 1909. Collections Zoologiques du baron Edm. de Sélys-Longchamps. Catalogue systématique et descriptif 8: 186 ▸ Indonesia: Timor

S. malayanus (MacLachlan, 1871). Journal of the Linnean Society of London. Zoology 11: 257 ▸ Indonesia: Java, Sulawesi; Philippines

S. salvazanus Navás, 1923. Memorie della Pontificia Accademia delle Scienze, Nuovi Lincei (2) 6: 4 ▸ Indochina

S. serratus Fraser, 1922. Records of the Indian Museum 24: 513 ▸ Thailand

Myrmeleontidae

ACANTHACLISINAE

Acanthaclisis Rambur, 1842
A. fuscata (Fabricius, 1781). Species Insectorum, exhibentes eorum differentias specificas, synonyma auctorum, loca natalia, metamorphosin adjectis observationibus: 386 ‣ E. Indies
**A. kawaii* Nakahara, 1913. Entomological News 24: 297 ‣ [Taiwan]

Heoclisis Navás, 1923
H. louiseae Banks, 1938. Journal of the Federated Malay States Museums 18: 224 ‣ Malaysia

Madrastra Navás, 1912
M. handlirschi Navás 1912. Memorias de la Real Academia de Ciencias y Artes de Barcelona 10: 8 ‣ Philippines

Stiphroneura Gerstaecker, [1885]
S. inclusus (Walker, 1853). Catalogue of the specimens of Neuropterous insects in the collection of the British Museum II: 327 ‣ E. Indies [India; Myanmar]

PALPARINAE

Echthromyrmex MacLachlan, [1867]
E. orientalis MacLachlan, 1873. Annales de la Société Entomologique Belge 16: 143 ‣ Indonesia: Moluccas [Myanmar; Thailand; India; Sri Lanka]

Palpares Rambur, 1842
P. pardus Rambur, 1842. Histoire Naturelle des insectes: Névroptères. Tome III: 375 ‣ E. Indies [India; Myanmar]
P. spectrum Rambur, 1842. Histoire Naturelle des insectes: Névroptères. Tome III: 376 ‣ 'E. Indies' [Africa]
**P. falcatus* MacLachlan, 1867. Journal of the Linnean Society of London. Zoology 9: 236 ‣ [Myanmar]
**P. nicholi* Navás, 1912. Memorias de la Real Academia de Ciencias y Artes de Barcelona 10: 6 ‣ [Myanmar]

Stenares Hagen, 1866
S. improbus (Walker, 1853). Catalogue of the specimens of Neuropterous insects in the collection of the British Museum II: 326 ‣ 'E. Indies' [India]

Tomatares Hagen, 1866

T. pardalis (Fabricius, 1781). Species Insectorum, exhibentes eorum differentias specificas, synonyma auctorum, loca natalia, metamorphosin adjectis observationibus: 398 ▸ Indonesia: Java [India]

MYRMELEONTINAE

Acratoleon Banks, [1915]

A. dispar Banks, 1934. Journal of the Federated Malay States Museums 17: 571 ▸ Borneo: Sabah

A. flavum Banks, 1915. Proceedings of the Academy of Natural Sciences of Philadelphia 66: 620 ▸ Papua New Guinea: Solomon Islands

Bandidus Navás, 1914

B. brandti New, 1990. Invertebrate Taxonomy 4: 14 ▸ Papua New Guinea

B. gressitti New, 1990. Invertebrate Taxonomy 4: 17 ▸ Papua New Guinea

B. sinewitensis New, 1990. Invertebrate Taxonomy 4: 16 ▸ Papua New Guinea

Brisus Navás, 1931

B. salomonicus Navás, 1931. Revista de la Academia de Ciencias Exactas, Físico-Químicas y Naturales de Madrid 26: 67 ▸ Papua New Guinea: Solomon Islands

Bullanga Navás, 1917

**B. binaria* Navás, 1917. Memorie della Pontificia Accademia delle Scienze, Nuovi Lincei (2) 3: 17 ▸ [Vietnam]

Cuca Navás, 1923

**C. impressa* Navás, 1923. Arxius de la secció de ciències. Institud d'estudis Catalans. Barcelona 7: 188 ▸ [Indochina]

Cueta Navás, 1911

**C. duplicata* Navás, 1914. Insecta : revue illustrée d'entomologie; publication mensuelle de la Station Entomologique de la Faculté des Sciences de Rennes 4: 138 ▸ [Vietnam]

**C. pilosa* Navás, 1917. Memorie della Pontificia Accademia delle Scienze, Nuovi Lincei 3: 14 ▸ [Vietnam]

**C. plicata* Navás, 1914. Insecta : revue illustrée d'entomologie; publication mensuelle de la Station Entomologique de la Faculté des Sciences de Rennes 4: 136 ▸ [Vietnam]

Cymothales Gerstaecker, [1894]

C. lestagei LaCroix, 1920. Bulletin de la Société entomologique de France 1920: 300 ▸ Borneo

Dendroleon Brauer, 1866

D. javanus Banks, 1914. Proceedings of the Academy of Natural Sciences of Philadelphia 66: 617 ▸ Indonesia: Java

D. kiungaensis New, 1990. Invertebrate Taxonomy 4: 12 ▸ Papua New Guinea

D. sumatranus Banks, 1918. Bulletin of the Museum of Comparative Zoology at Harvard 62: 16 ▸ Indonesia: Sumatra

D. vitripennis (Navás, 1912). Memorias de la Real Academia de Ciencias y Artes de Barcelona 10: 39 ▸ Malaysia

Distoleon Banks, 1910

D. bakeri Banks, 1916. The Philippine journal of science 11: 210 ▸ Philippines

D. bistrigatus (Rambur, 1842). Histoire Naturelle des insectes: Névroptères. Tome III: 391 ▸ New Guinea, [Australia]

D. somnolentus (Gerstaecker, 1884). Mitteilungen aus dem Naturwissenschaftlichen Verein für Neu-Vorpommern und Rügen in Greifswald 16: 13 ▸ New Guinea, [Australia]

Dolicholeon Navás, [1929]

**D. burmanus* Navás, 1931. Revista de la Academia de Ciencias Exactas, Físico-Químicas y Naturales de Madrid 26: 80 ▸ [Myanmar]

Eophanes Banks, 1931

E. formosa Banks, 1931. Psyche 38: 61 ▸ Philippines

Episalus Gerstaecker, [1885]

E. zephyrinus Gerstaecker, 1884. Mitteilungen aus dem Naturwissenschaftlichen Verein für Neu-Vorpommern und Rügen in Greifswald 16: 19 ▸ Papua New Guinea

Feinerus Navás, 1919

**F. formosanus* Navás, 1926. Memorie della Pontificia Accademia delle Scienze, Nuovi Lincei (2) 9: 112 ▸ [Taiwan]

**F. umbratus* Navás, 1919. Insecta : revue illustrée d'entomologie; publication mensuelle de la Station Entomologique de la Faculté des Sciences de Rennes 9: 191 ▸ [Vietnam]

Formicaleon Banks, 1911

F. cleonice Banks, 1914. Proceedings of the Entomological Society of Washington 15: 172 ▸ Philippines

**F. feai* Navás, 1915. Memorias de la Real Academia de Ciencias y Artes de Barcelona 11: 19 ▸ [Myanmar]

**F. formosanus* Okamoto, 1910. Wiener Entomologische Zeitung 29: 291 ▸ [Taiwan]

F. latipennis Navás, 1914. Annales de la Société Scientifique de Bruxelles 38: 253 ▸ Indonesia: Sulawesi

F. masi Navás, 1916. Memorie della Pontificia Accademia delle Scienze, Nuovi Lincei (2) 4: 5 ▸ Philippines

F. schaedleri van der Weele, 1909. Notes from the Leyden Museum 31: 23 ▸ Indonesia: Timor

F. tripunctatus Navás, 1914. Insecta : revue illustrée d'entomologie; publication mensuelle de la Station Entomologique de la Faculté des Sciences de Rennes 4: 139 ▸ Thailand [Cambodia]

Hagenomyia Banks, 1911

**H. asakurae* (Okamoto, 1910). Wiener Entomologische Zeitung 29: 297 ▸ [Taiwan]

**H. brunneipennis* Esben-Petersen, 1913. Entomologische Mitteilungen 2: 223 ▸ [Taiwan]

H. sagax (Walker, 1853). Catalogue of the specimens of Neuropterous insects in the collection of the British Museum II: 382 ▸ Indonesia; Papua New Guinea

H. sumatranus (van der Weele, 1909). Notes from the Leyden Museum 31: 34 ▸ Indonesia [Sri Lanka]

Indoleon Banks, 1913

I. tacitus (Walker, 1853). Catalogue of the specimens of Neuropterous insects in the collection of the British Museum II: 362 ▸ 'E. Ind.'; Malaysia

Mossega Navás, [1914]

M. papuensis New, 1990. Invertebrate Taxonomy 4: 10 ▸ Papua New Guinea

M. waidoraensis New, 1990. Invertebrate Taxonomy 4: 10 ▸ Papua New Guinea

Myrmeleon Linnaeus, 1767

M. acer Walker, 1853. Catalogue of the specimens of Neuropterous insects in the collection of the British Museum II: 348 ▸ Indonesia: Timor, Irian Jaya; Papua New Guinea, [Australia]

M. angustipennis Banks, 1916. The Philippine journal of science 11: 209 ▸ Philippines

M. capito Navás, 1912. Mitteilungen der Münchener Entomologischen Gesellschaft 3: 90 ▸ New Guinea

M. celebensis MacLachlan, 1875. Tijdschrift voor Entomologie 18: 5 ▸ Malaysia; Indonesia

M. ermineus Fabricius, 1798. Supplementum Entomologiae systematicae: 206 ▸ 'E. Indies'

M. frontalis Burmeister, 1839. Handbuch der Entomologie 2: 993 ▸ Indonesia: Java, Timor

M. iridescens Kirby, 1900. In: Andrews, C.A. (ed.) A Monograph of Christmas Island, Indian Ocean, London: 140 ▸ Australia: Christmas Island

**M. krempfi* Navás, 1923. Arxius de la secció de ciències. Institud d'estudis Catalans. Barcelona 7: 182 ▸ [Vietnam]

M. pictifrons Gerstaecker, 1885. Mitteilungen aus dem Naturwissenschaftlichen Verein für Neu-Vorpommern und Rügen in Greifswald 16: 96 ▸ Papua New Guinea [Australia]

M. punctatus Fabricius, 1787. Mantissa Insectorum sistens eorum species nuper detectas: 249 ▸ 'E. Indies'

M. ursinus Fabricius, 1798. Supplementum Entomologiae systematicae: 207 ▸ 'E. Indies'

M. viganus Navás, 1923. Memorie della Pontificia Accademia delle Scienze, Nuovi Lincei (2) 6: 37 ▸ Philippines

Neleinus Navás, 1915

**N. turbatus* Navás, 1915. Memorias de la Real Academia de Ciencias y Artes de Barcelona 11: 13 ▸ [Myanmar]

Nepsalus Navás, [1914]

N. indicus Navás, 1914. Annales de la Société Scientifique de Bruxelles 38: 251 ▸ W. Malaysia; Borneo; Indonesia: Sulawesi

Neurocolinus Navás, 1930

N. philippinus (Navás, 1925). Memorias de la Real Academia de Ciencias y Artes de Barcelona 19: 9 ▸ Philippines

Neuroleon Navás, 1909

N. decoratum Handschin, 1937. Verhandlungen der Naturforschenden Gesellschaft in Basel 47: 127 ▸ Indonesia: Timor

N. disjunctus Banks, 1913. Proceedings of the Entomological Society of Washington 15: 173 ▸ Malaysia: Sabah; Philippines

N. indistinctum Navás, 1914. Memorias de la Real Academia de Ciencias y Artes de Barcelona 11: 303 ▸ Malaysia

N. *uniformis* Navás, 1914. Revue russe d'entomologie 14: 7 ▸ Malaysia

Nomes Navás, 1914

N. *lorianus* Navás, 1914. Revista de la Academia de Ciencias Exactas, Físico-Químicas y Naturales de Madrid 13: 234 ▸ New Guinea

Nuglerus Navás, 1912

N. *maculata* (Navás, 1915). Memorias de la Real Academia de Ciencias y Artes de Barcelona 11: 14 (as *Bofia*) ▸ W. Malaysia; Thailand ['Indochina']

N. *scalaris* Navás, 1912. Annales de la Société Scientifique de Bruxelles 36: 245 ▸ Malaysia [N. India]

Paraglenurus van der Weele, 1909

P. *borneensis* van der Weele, 1909. Notes from the Leyden Museum 31: 30 ▸ Borneo

P. *scopifer* (Gerstaecker, 1888). Mitteilungen aus dem Naturwissenschaftlichen Verein für Neu-Vorpommern und Rügen in Greifswald 19: 110 ▸ Indonesia: Ceram

Periclystus Gerstaecker, [1888]

P. *vicinus* New, 1985. Journal of the Australian Entomological Society 24: 229 ▸ Papua New Guinea

Phanoleon Banks, 1931

P. *bicostatus* Banks, 1931. Journal of the Federated Malay States Museums 16: 387 ▸ Malaysia

Pseudoformicaleo van der Weele, 1909

*P. *barbatus* Navás, 1923. Arxius de la secció de ciències. Institud d'estudis Catalans. Barcelona 7: 189 ▸ [Indochina]

P. *jacobsoni* van der Weele, 1909. Notes from the Leyden Museum 31: 25 ▸ Indonesia

P. *nubecula* (Gerstaecker, 1885). Mitteilungen aus dem Naturwissenschaftlichen Verein für Neu-Vorpommern und Rügen in Greifswald 16: 101. ▸ Indonesia; Malaysia; Papua New Guinea [Taiwan; Japan; China; Australia]

Salvaza Navás, 1917

**S. cornutus* Navás, 1917. Insecta : revue illustrée d'entomologie; publication mensuelle de la Station Entomologique de la Faculté des Sciences de Rennes 7: 12 ▸ [Vietnam]

Stenogymnocnemia Esben-Petersen, 1923

S. ismayi New, 1990. Invertebrate Taxonomy 4, 18 ▸ Papua New Guinea

References

A few references not directly referred to in the text but important in providing additional basic information are also included and indicated by an asterisk (*).

Ackery, P.R. & Vane-Wright, R.I., 1984. Milkweed butterflies. British Museum (Natural History), London: vii + 425 pp.

Adams, P.A., 1970. A review of the New World Dilaridae.– Postilla 141: 1-11.

Adams, P.A. & Penny, N.D., 1987. Neuroptera of the Amazon Basin part 11a. Introduction and Chrysopini.– Acta Amazonica 15: 413-479 (1985).

Aiken, S.R., 1994. Peninsular Malaysia's protected areas coverage, 1903-92: creation, rescission, excision and intrusion.– Environmental Conservation 21: 49-56.

Aspöck, U., 1983. Das genus *Berotha* Walker (Neuropteroidea: Planipennia: Berothidae).– Annalen des Naturhistorischen Museums Wien 84: 463-478.

Aspöck, U., 1986. The present state of knowledge of the family Berothidae (Neuropteroidea: Planipennia).– In: Gepp, J., Aspöck, H. & Hölzel, H. (eds). Recent research in Neuropterology, Graz: 87-101.

Aspöck, U.,1992. Crucial points in the phylogeny of the Neuroptera (Insecta).– In: Canard, M., Aspöck, H. & Mansell, M.W. (eds). Current research in Neuropterology. Toulouse: 63-73.

***Aspöck**, U. & Aspöck, H., 1983. Das Genus *Isoscelipteron* Costa 1863 (Neuropteroidea: Planipennia: Berothidae).– Zeitschrift der Arbeitsgemeinschaft Osterreichischer Entomologen 32: 65-74.

Aspöck, U. & Aspöck, H., 1991. Zur Kenntnis des Genus *Isoscelipteron* Costa 1863 (Neuropteroidea: Neuroptera: Berothidae: Berothinae).– Zeitschrift der Arbeitsgemeinschaft Österreichischer Entomologen 43: 65-76.

Aspöck, U. & Aspöck, H., 1999. Kamelhälse, Schlammfliegen, Ameisenlöwen. Wer sind see? (Insecta: Neuropterida: Raphidioptera, Megaloptera, Neuroptera). Stapfia 60: 1-34.

Aspöck, H. & Mansell, M.W., 1994. A revision of the family Rhachiberothidae Tjeder, 1959, stat. n. (Neuroptera).– Systematic Entomology 19: 181-206.

Aspöck, U. & Nemeschkal, N.L., 1998. A cladistic analysis of the Berothidae (Neuroptera).– Acta zoologica Fennica 209: 45-63.

Aspöck, U., Plant, J.D. & Nemeschkal, H.L., 2001. Cladistic analysis of Neuroptera and their systematic position within Neuropteroidea (Insexta: Neuropterida: Neuroptera).– Systematic Entomology 26: 73-86.

Banks, N., 1899. A classification of the North American Myrmeleonidae.– Canadian Entomologist 31: 67-71.

Banks, N., 1911. Notes on African Myrmeleonidae.– Annals of the Entomological Society of America 4: 1-29.

***Banks**, N., 1915. New neuropteroid insects, native and exotic.– Proceedings of the Academy of Natural Sciences, Philadelphia 66: 608-632.

Banks, N., 1929. Fauna of the Batu caves, Selangor. xiii. Neuroptera. On the myrmeleonid *Neglurus vitripennis* Navás.– Journal of the Federated Malay States Museums xiv: 372-373.

Barlow, H.S., 1988. Forest Lepidoptera.– In: Cranbrook, G. G.-H., Earl of (ed.). Key environments. Malaysia. Pergamon Press, Oxford: 212-224.

Barnard, P.C, 1981. The Rapismatidae (Neuroptera): montane lacewings of the Oriental region.– Systematic Entomology 6: 121-136.

Barnard, P.C. & New, T.R., 1985. New species in the *Rapisma malayanum*-complex (Neuroptera: Rapismatidae).– Neuroptera International 3: 165-173.

Braatz, S., 1992. Conserving biological diversity: a strategy for protected areas in the Asia-Pacific region.– World Bank Technical Paper no. 193. Washington DC: xiii + 66 pp.

Brooks, S.J., 1983. A new genus of oriental lacewings (Neuroptera: Chrysopidae).– Bulletin of the British Museum of Natural History (Entomology) 47: 1-26.

Brooks, S.J., 1994. A taxonomic review of the common green lacewing genus *Chrysoperla* (Neuroptera: Chrysopidae).– Bulletin of the Natural History Museum, London (Entomology) 63: 137-210.

Brooks, S.J., 1997. An overview of the current status of Chrysopidae (Neuroptera) systematics.– Deutsche Entomologische Zeitung 44: 267-275.

Brooks, S.J. & Barnard, P.C., 1990. The green lacewings of the world: a generic review (Neuroptera: Chrysopidae).– Bulletin of the British Museum of Natural History (Entomology) 59: 117-286.

Canard, M., Aspöck, H. & Mansell, M.W., 1992. Nomenclatural note: Myrmeleontidae vs Myrmeleonidae.– In: Canard, M., Aspöck, H. & Mansell, M W. (eds). Current research in Neuropterology, Toulouse: 253-254.

Canard, M., Séméria, Y. & New, T.R. (eds) 1984. Biology of Chrysopidae. W. Junk, Dordrecht: x + 294 pp.

Dammerman, K. W., 1929. The agricultural zoology of the Malay peninsula. J.H. de Bussy, Amsterdam: xi + 473 pp.

Dickersen, R.E., Merrill, E.D., McGregor, R.C., Schultze, W., Taylor, E.H. & Herre, A.W.C.T., 1928. Distribution of life in the Philippines. Manila. Bureau of Sciences Monograph 21, 1-322. (Reprint 1975, Bookmark, Manila).

Duffels, J.P. & de Boer, A.J., 1990. Areas of endemism and composite areas in East Malesia.– In: Baas, P., Kalkman, K. & Geesink, R. (eds). The plant diversity of Malesia. Kluwer Academic Publishers, Dordrecht: 249-272.

Enderlein, G., 1910. Klassifikation der Mantispiden nach dem Material des Stettiner Zoologischen Museums.–. Stettiner Entomologische Zeitung 71: 341-379.

Esben-Petersen, P., 1918. Help-notes towards the determination and the classification of the European Myrmeleonidae.– Entomologische Meddelelser 12: 97-127.

Esben-Petersen, P., 1933. Neuroptera der Deutschen Limnologischen Sunda-Expedition.– Archive fur Hydrobiologie, Supplement 11: 625-630.

Ghilarov,M.S., 1962. The larva of *Dilar turcicus* Hag. and the position of the family Dilaridae in the order Planipennia. - Entomologicheskoye Obozreniye, Moscow 41: 402-416.

Ghosh, S.K., 1988. Contribution to the taxonomical studies of Neuroptera (suborder Planipennia) from eastern India. II. Family Ascalaphidae.– Records of the Zoological Survey of India 85: 163-191.

Handschin, E., 1937. Neuroptera von Timor und Rote.– Verhandlungen des Naturforschende Gesellschaft Basel 47: 123-135.

Handschin, E., 1961. Beitrage zur Kenntnis der gattungen *Euclimacia, Climaciella* und *Entanoneura* Enderlein 1910 in Indo-Australischen Faunengebiet.– Nova Guinea (Zoology) 15: 253-301.

Henry, C.S., 1972. Eggs and repagula of *Ululodes* and *Ascalaptynx* (Neuroptera: Ascalaphidae): a comparative study.– Psyche, Cambridge Mass. 79: 1-22.

***Henry**, C.S., 1976. Some aspects of the external morphology of larval owlflies (Neuroptera: Ascalaphidae), with particular reference to *Ululodes* and *Ascaloptynx*.– Psyche 83: 1-31.

Hölzel, H., 1972. Die Neuropteren Vorderasiens. iv. Myrmeleonidae.– Beitrage zur Naturkundlichen Forschung in Südwestdeutschland 1: 3-103.

Hölzel, H., 1986. Biogeography of Palaearctic Myrmeleonidae (Neuropteroidea: Planipennia).– In: Gepp, J., Aspöck, H & Hölzel, H. (eds). Recent research in Neuropterology. Graz: 53-70.

Huxley, T.H., 1868. On the classification and distribution of the Alecteromorphae and Heteromorphae. – Proceedings of the Zoological Society of London (1868): 294-319.
Johnson, J.B. & Hagen, K. S., 1981. A neuropterous larva uses an allomone to attack termites.– Nature 289: 506-507.
Kato, M., 1989. The fern flora of Seram.– In: Baas, P., Kalkman, K. & Geesink, R. (eds). The plant diversity of Malesia. Kluwer Academic Publishers, Dordrecht: 225-234.
Kiew, B.-H., 1970a. The ecology of *Myrmeleon celebensis* McLachlan (Neuroptera: Myrmeleontidae).– Malayan Nature Journal 24: 11-15.
Kiew, B.-H., 1970b. Life history of *Myrmeleon celebensis* McLachlan (Neuroptera: Myrmeleontidae).– Indian Journal of Entomology 32: 230-245.
Kimmins, D.E., 1942 The genus *Thyridosmylus* Krüger, with notes on the subfamily Spilosmylinae (Neuroptera, Osmylidae).– Annals and Magazine of Natural History (11) 9: 848-855.
Kimmins, D.E., 1957. 16. Odonata, Neuroptera and Trichoptera.– Natural history of Rennell Island, British Solomon Islands 2: 67-70.
Krüger, L., 1912. Osmylidae. Beiträge zu einer Monographie der Neuropteren-Familie der Osmyliden.– Stettiner Entomologische Zeitung 73: 319-373.
Krüger, L., 1913a. Osmylidae. Beiträge zu einer Monographie der Neuropteren-Familie der Osmyliden II.– Stettiner Entomologische Zeitung 74: 3-123.
Krüger, L., 1913b. Osmylidae. Beiträge zu einer Monographie der Neuropteren familie der Osmyliden III-IV.– Stettiner Entomologische Zeitung 74: 193-294.
Krüger, L., 1914. Osmylidae. Beiträge zu einer Monographie der Neuropteren familie der Osmyliden V.– Stettiner Entomologische Zeitung 75: 9-130.
Krüger, L., 1915. Osmylidae. Beiträge zu einer Monographie der Neuropteren familie der Osmyliden VI.– Stettiner Entomologische Zeitung 76: 60-87.
Kuwayama, S., 1962. A revisional synopsis of the Neuroptera in Japan.– Pacific Insects 4: 325-412.
Lambkin, K.L., 1986a. A revision of the Australian Mantispidae (Insecta: Neuroptera) with a contribution to the classification of the family. I. General and Drepanicinae.– Australian Journal of Zoology, Supplement 116: 1-142.
Lambkin, K.L., 1986b. A revision of the Australian Mantispidae (Insecta: Neuroptera) with a contribution to the classification of the family. II. Calomantispinae and Mantispinae. – Australian Journal of Zoology, Supplement 117: 1-112.
Laumonier, Y., 1997. The vegetation and physiography of Sumatra. Kluwer Academic Publishers, Dordrecht: ix + 222 pp.
MacKinnon, J. (ed.) 1997. Protected areas systems review of the Indo-Malayan realm. Asian Bureau for Conservation Ltd, Hong Kong/World Conservation Monitoring Centre, Cambridge: x + 198 pp.
MacKinnon, J. & MacKinnon, K., 1986. Review of the protected areas system in the Indo Malayan realm. IUCN, Gland: xii + 284 pp.
MacLeod, E.G. & Adams, P.A., 1967. A review of the taxonomy and morphology of the Berothidae, with the description of a new subfamily from Chile (Neuroptera).– Psyche 74: 237-265.
MacLeod, E.G. & Spiegler, P.E., 1961. Notes on the larval habitat and developmental peculiarities of *Nallachius americanus* (MacLachlan; Neuroptera: Dilaridae).– Proceedings of the Entomological Society of Washington 63: 281-286.
Makarkin, B.H., 1985. [Review of the Osmylidae of the USSR.] (in Russian). In: Lev, P.A. & Storozhenko, S. (eds). Taksonomiia i ekologiia chlenistoogikh Dal'nego Vostoka: sbornik nauchnykh trudov. Dalnevostochnyi Nauchnyi Tsentre AN SSR,Vladivostok: 35-51.
Mansell, M.W., 1985. The antlions of southern Africa (Neuroptera, Myrmeleontidae). Introduction and genus *Bankisus* Navás. Journal of the Entomological Society of southern Africa 48: 189-212.

Mansell, M.W., 1987. The antlions of southern Africa (Neuroptera: Myrmeleontidae): genus *Cymothales* Gerstaecker, including extralimital species.– Systematic Entomology 12: 181-219.

Mansell, M.W., 1990. The Myrmeleontidae of southern Africa: tribe Palparini. Introduction and description of *Pamares* gen. nov. with four new species (Insecta: Neuroptera).– Journal of the Entomological Society of southern Africa 53: 165-189.

Mansell, M.W., 1992. Key characters in the phylogeny and classification of Palparini (Insecta: Neuroptera: Myrmeleontidae). In: Canard, M., Aspöck, H. & Mansell, M.W. (eds). Current research in Neuropterology, Toulouse: 243-254.

Mansell, M.W., 1999. Evolution and success of antlions (Neuropterida: Neuroptera: Myrmeleontidae).– Stapfia 60 (no. 138): 49-58.

Markl, W., 1954. Vergleichend-morphologische Studien zur Systematik und Klassifikation der Myrmeleoniden.– Verhandlungen der Naturforschenden Gesellschaft in Basel 65: 178-263.

McEwen, P.K., New, T.R. & Whittington, A.E. (eds) 2001. Lacewings in the crop environment. Cambridge University Press, Cambridge, New York: xviii + 546 pp.

Meinander, M., 1972. A revision of the family Coniopterygidae (Planipennia).– Acta zoologica Fennica 136: 1-357.

Meinander, M., 1990. The Coniopterygidae (Neuroptera, Planipennia). A check-list of the species of the world, descriptions of new species, and other new data.– Acta zoologica Fennica 189: 1-95.

Monserrat, V.J., 1977. A systematic and alphabetic list of Neurorthidae and Sisyridae (Neuroptera).– Nouvelle Revue d'Entomologie 7: 91-96.

Monserrat, V.J., 1981. Sobre los Sisíridos de la Región Oriental (Neuroptera, Planipennia, Sisyridae).– Eos 57: 165-186.

Monserrat, V.J., 1982. Contribución al conocimiento de los coniopterígidos (Insecta, Neuroptera, Coniopterygidae) de la región oriental.– Annali del Museo Civico di Storia Naturale Giacomo Doria 84: 9-39.

***Monserrat**, V.J., 1990. A systematic checklist of the Hemerobiidae of the world (Insecta: Neuroptera).– In: Mansell, M.W. & Aspöck, H. (eds). Advances in Neuropterology. Pretoria: 215-262.

Murphy, D.H. & Lee, Y.T., 1971. Three new species of *Coniopteryx* from Singapore (Planipennia: Coniopterygidae).– Journal of Entomology B 40: 151-161.

***Nakahara**, W., 1955. The Dilaridae of Japan and Formosa (Neuroptera).– Kontyû 23: 133-142.

Nakahara, W., 1960. Systematic studies on the Hemerobiidae (Neuroptera).– Mushi 34: 1-69.

Navás, L., 1913. Neuroptera asiatica. II.– Revue Russe d'Entomologie 13: 424-430.

Navás, L., 1919. Algunos Neurópteros de Argentina. Revista de la real Academia de Ciencias exactas, físicas y naturales de Madrid 17: 287-305.

Navás, L., 1923. Insecta nova. xii. Memorie della Pontificia Accademia delle Scienze, Nuovi Lincei 6: 1-8.

***Navás**, L., 1929. Monografía de la familia de los Berótidos (Insecta, Neuroptera).– Memorias de la real Academia de Ciencias exactas, físico-químicas y naturales de Zaragoza 2: 1-107.

Navás, L., 1931. Decadas de Insectos nuevos. Revista de la real Academia de Ciencias exactas, físicas y naturales de Madrid 26: 60-86.

New, T.R., 1981. A revision of the Australian Nymphidae (Insecta: Neuroptera).– Australian Journal of Zoology 29: 707-750.

New, T.R., 1982. A reappraisal of the status of the Stilbopterygidae (Neuroptera: Myrmeleontoidea).– Journal of the Australian Entomological Society 21: 71-75.

New, T.R., 1983. Some early stages of *Osmylops* (Neuroptera: Nymphidae).– Systematic Entomology 8: 121-126.

New, T.R., 1984. Revision of the Australian Ascalaphidae (Insecta: Neuroptera).– Australian Journal of Zoology, Supplement 100: 1-86.

New, T.R., 1985a. A revision of the Australian Myrmeleontidae (Insecta: Neuroptera). I. Introduction, Myrmeleontini, Protoplectrini.– Australian Journal of Zoology, Supplement 104: 1-90.

New, T.R., 1985b. A revision of the Australian Myrmeleontidae (Insecta: Neuroptera). II. Dendroleontini.– Australian Journal of Zoology, Supplement 105: 1-170.

New, T.R., 1985c. A revision of the Australian Myrmeleontidae (Insecta: Neuroptera). III. Distoleontini and Acanthaclisinae.– Australian Journal of Zoology, Supplement 106: 1-159.

New, T.R., 1985d. A new species of *Periclystus* Gerstaecker (Neuroptera: Myrmeleontidae) from Papua New Guinea.– Journal of the Australian Entomological Society 24: 229-231.

*__New__, T.R., 1985e. A new species of Rapismatidae (Neuroptera) from Sabah.– Neuroptera International 3: 133-135.

*__New__, T.R., 1986a. A review of the biology of Neuroptera Planipennia.– Neuroptera International, Supplement 1: 1-56.

New, T.R., 1986b. Osmylidae (Insecta: Neuroptera) from New Guinea.– Australian Journal of Zoology 34: 855-884.

New, T.R., 1988a. A revision of the Australian Hemerobiidae (Insecta: Neuroptera).– Invertebrate Taxonomy 2: 339-411.

New, T.R., 1988b. Osmylidae (Insecta: Neuroptera) from Irian Jaya.– Invertebrate Taxonomy 2: 805-839.

New, T.R., 1989a. Planipennia. Lacewings. Handbuch der Zoologie Vol. 4 Arthropoda: Insecta. W. de Gruyter, Berlin: xi + 132 pp.

New, T.R., 1989b. Hemerobiidae (Insecta: Neuroptera) from New Guinea.– Invertebrate Taxonomy 2 (1988): 605-632.

New, T.R., 1989c. Preliminary appraisal of a tropical lacewing fauna: local diversity of Neuroptera around Wau, Papua New Guinea.– Neuroptera International 5: 211-218.

New, T.R., 1990a. Myrmeleontidae (Insecta: Neuroptera) from New Guinea.– Invertebrate Taxonomy 4: 1-20.

New, T.R., 1990b. The Neuroptera of Cape York Peninsula, northern Queensland: a Torresian fauna in Australia?– Neuroptera International 6: 9-17.

*__New__, T.R., 1990c. The genus *Heteroconis* Enderlein (Insecta: Neuroptera: Coniopterygidae) in New Guinea.– Invertebrate Taxonomy 4: 665-684.

New, T.R., 1991a. Osmylidae (Insecta: Neuroptera) from the Oriental Region.– Invertebrate Taxonomy 5: 1-31.

New, T.R, 1991b. Neuroptera from Christmas Island, Indian Ocean.– Neuroptera International 6: 133-136.

New, T.R., 1992. The lacewings (Insecta, Neuroptera) of Tasmania. – Papers and Proceedings of the Royal Society of Tasmania 126: 29-45.

New, T.R., 1996. Neuroptera. – In: Wells, A. (ed.). Zoological Catalogue of Australia. Vol. 28. CSIRO Publishing, Melbourne: 1-104, 184, 199-216.

New, T.R., 1997. Neuroptera of Australia: Faunal elements, diversity and relationships.– Deutsche Entomologische Zeitung 44: 259-265.

New, T.R., 1998. Preliminary survey of the Mantispidae (Neuroptera) of southeast Asia.– Acta zoologica Fennica 209: 175-181.

New, T.R. & Sudarman, H.K., 1988. Neuroptera (Insecta) of the Krakatau Islands, Indonesia.– Philosophical Transactions of the Royal Society of London, B 322: 413-426.

Newman, E., 1853. A proposed division of the Neuroptera into two classes.– Zoologist 11. Appendix: 181-204.

Oswald, J.D., 1993a. Revision and cladistic analysis of the world genera of the family Hemerobiidae (Insecta: Neuroptera).– Journal of the New York Entomological Society 101: 143-299.

Oswald, J.D., 1993b. Phylogeny, taxonomy and biogeography of extant silky lacewings (Insecta: Neuroptera: Psychopsidae).– Memoirs of the American Entomological Society 40: 1-65.

Oswald, J.D., 1994. A new phylogenetically basal subfamily of brown lacewings from Chile (Neuroptera: Hemerobiidae).– Entomologica Scandinavica 25: 295-302.

Oswald, J.D., 1997. Review of the *sejunctus* group of the split-footed lacewing genus *Osmylops* Banks (Neuroptera: Nymphidae) with remarks on the functional morphology of terminalic coupling.– Australian Journal of Entomology 36: 351-358.

Oswald, J.D., 1998a. *Osmylops* Banks (Neuroptera: Nymphidae): generic review and revision of the *armatus* species group.– Journal of Neuropterology 1: 79-108.

Oswald, J.D., 1998b. Annotated catalogue of the Dilaridae (Insecta: Neuroptera) of the world.– Tijdschrift voor Entomologie 141: 115-128.

Oswald, J.D. & Schiff, N.M., 2001. A new species of the genus *Dilar* Rambur (Neuroptera: Dilaridae) from Borneo. – Proceedings of the Entomological Society of Washington 103: 74-80.

Parfin, S. & Gurney, A.B., 1956. The spongilla-flies, with special references to those of the western hemisphere (Sisyridae, Neuroptera).– Proceedings of the United States National Museum 105: 421-529.

Pupedis, R., 1987. Foraging behavior and food of adult spongilla-flies (Neuroptera: Sisyridae).– Annals of the Entomological Society of America 80: 758-760.

Redborg, K. & MacLeod, E.G., 1984. The developmental ecology of *Mantispa uhleri* Banks (Neuroptera, Mantispidae).– Illinois Biological Monographs 3: 1-130.

Riek, E.F.,1970. Neuroptera (lacewings) .pp. 472-494 in Mackerras, I.M. (ed.) The insects of Australia. Melbourne University Press, Melbourne.

Simpson, G.G., 1977. Too many lines: the limits of the Oriental and Australian zoogeographic regions.– Proceedings of the American Philosophical Society 121: 107-120.

Stange, L.A., 1970. Revision of the antlion tribe Brachynemurini of North America.– University of California Publications in Entomology 55: 1-192.

Stange, L.A., 1976. Classificación y catálogo mundial de la tribu Dendroleontini con la redescripción de género *Voltor* Navás (Neuroptera: Myrmeleontidae).– Acta Zoológica Lilloana 31:261-322.

Stange, L.A., 1989. Review of the new World Dimarini with the description of a new genus from Peru (Neuroptera: Myrmeleontidae).– Florida Entomologist 72: 450-461.

Stange, L.A. & Miller, R.B., 1985. A generic review of the Acanthaclisine antlions based on larvae (Neuroptera: Myrmeleontidae). Insecta Mundi 1: 29-42.

Stange, L.A. & Miller, R.B., 1990. Classification of the Myrmeleontidae based on larvae (Insecta: Neuroptera). In: Mansell, M.W. & Aspöck, H. (eds). Advances in Neuropterology, Pretoria: 151-169.

Stange, L.A. & Wang, H.Y., 1998. Guide book to insects of Taiwan (18): Neuroptera - Megaloptera - Raphidioptera. Taipei: 279 pp.

Sziraki, G., 1999. An annotated list of the Ascalaphidae species known from Asia and from the Pacific Islands.– Folia Entomologica Hungarica 59: 57-72.

Tjeder, B., 1957. Neuroptera - Planipennia. The lacewings of southern Africa. 1. Introduction and families Coniopterygidae, Sisyridae and Osmylidae.– South African Animal Life 4: 95-188.

Tjeder, B, 1959. Neuroptera - Planipennia. The lacewings of southern Africa. 2. Family Berothidae.– South African Animal Life 6: 256-314.

Tjeder, B., 1961. Neuroptera - Planipennia. The lacewings of southern Africa. 4. Family Hemerobiidae.– South African Animal Life 8: 296-408.

Tjeder, B., 1966. Neuroptera - Planipennia. The lacewings of southern Africa. 5. Family Chrysopidae.– South African Animal Life 12: 228-534.

Tjeder, B., 1972. Two necessary alterations in long-established genus nomenclature in Ascalaphidae (Neuroptera).– Entomologica Scandinavica 3: 153-155.

Tjeder, B., 1992. The Ascalaphidae of the Afrotropical Region (Neuroptera). 1. External morphology and bionomics of the family Ascalaphidae, and taxonomy of the subfamily Haplogleniinae including the tribes Proctolyrini n. tribe, Melambrotini n. tribe, Campylophlebini n. tribe, Tmesibasini n. tribe, Allocormodini n. tribe and Ululomyiinae n. tribe.– Entomologica Scandinavica, Supplement 41: 3-169.

Tjeder, B. & Hansson, C. 1992. The Ascalaphidae of the Afrotropical region (Neuroptera). 2. Revision of the tribe Ascalaphini (subfamily Ascalaphinae) exluding the genus *Ascalaphus* Fabricius. Entomologica Scandinavica, Supplement 41: 171-237.

Tsukaguchi, S., 1985. A checklist of published species of Japanese Chrysopidae (Neuroptera).– Kontyû 53: 503-506.

Tsukaguchi, S., 1995. Chrysopidae of Japan (Insecta: Neuroptera). Privately published, Osaka: 223 pp.

Tsukagushi, S. & Yukawa, J., 1988. Neuroptera collected from the Krakatau Islands, Indonesia.– Kontyû 56: 481-490.

Turner, B.D., 1992. The colonisation of Anak Krakatau: interactions between wild sugar cane, *Saccharum spontaneum*, and the antlion *Myrmeleon frontalis*.– Journal of Tropical Ecology 8: 435-449.

van der Weele, H.W., 1909a. Mecoptera and Planipennia of Insulinde, with biological notes from Edw. Jacobson.– Notes from the Leyden Museum 31: 1-100.

van der Weele, H.W., 1909b. Ascalaphiden monographisch bearbeitet. Collections Zoologiques du baron Edm. de Selys-Longchamps. Catalogue systématique et descriptif 8: ii + 326 pp.

van Steenis, C.G.G.J., 1950. The delimitation of Malesia and its main plant geographical divisions.– Flora Malesiana 1. 1: lXX-lXXV.

van Steenis, C.G.G.J., 1972. The mountain flora of Java. Brill, Leiden: viii + 147 pp. [Illustrated by H. Amir & M. Toha.]

Wheeler, W.M., 1930. Demons of the dust. London, Trubner & Co.: xviii + 378 pp.

***Whitmore**, T.C. (ed.) 1981. Wallace's line and plate tectonics. Clarendon Press, Oxford: x + 91 pp.

Whitmore, T.C. (ed.) 1987. Biogeographical evolution of the Malay Archipelago. Clarendon Press, Oxford: x + 145 pp.

***Zakharenko**, A.V., 1991. Two new species of the family Dilaridae (Neuroptera) from Vietnam. Zoologicheskii Zhurnal 70: 142-144.

Glossary

This glossary explains some of the morphological terms used in keys and descriptions of taxa of Neuroptera. Some terms are very general, while others reflect the specialised terminology developed especially for Neuroptera.

a *abdominal gills* — In larvae of Sisyridae: the paired ventral articulated structures on the ventral side of the abdomen, used in respiratory exchange in freshwater habitat (fig. 36).

accessory sclerites — In male genitalia: refers to detached sclerites (single or paired) with uncertain homologies, and subsidiary to the main defined genitalic structures.

aedeagus — The male intromittent organ, used for copulation.

apodeme — A rigid sclerotised process that forms part of the insect endoskeleton or a base for attachment of muscles.

arcessus — In male genitalia: a structure attached ventrally to the upper median part of the gonarcus (fig. 84). The attachment is flexible, so that the structure is movable.

arolium — A median projection from the end of the pretarsus, between the two pretarsal claws (fig. 3), see also 'empodium'.

b *Banksian line* — In wings of some Myrmeleontidae: prominent longitudinal line(s) on one or both wings (fig. 1, 184), formed by bending and alignment of the branches of Rs (anterior Banksian line) or of CuA (forewing) or MP2 (hindwing) (posterior Banksian line). Either or both lines may be present, but both are commonly absent. The arrangement is of taxonomic value.

biareolate — In wing venation: a double row of closed cells (commonly in the costal area as in fig. 184).

bulla — A swollen or expanded area of the wing membrane, often slightly domed or otherwise differentiated in texture from the rest of the membrane (see embossed spot).

c *campodeiform* — A form of endopterygote insect larva, characterised by being elongate, prognathous and having long legs; typified by many active predatory larvae (figs. 81, 107).

cercal callus — A defined area or field on the ectoproct, bearing trichobothria (fig. 110).

coxa — The basal section of the insect leg, attached to the lateroventral region of the wall of each thoracic segment .

coxopodite — The primary basal segment of an arthropod appendage, representing the limb base; in legs, the coxa.

crumena — In Mantispidae: an anteriorly excavate median pouchlike structure on abdominal sternite vii of females of *Theristria*, where its position and size are of taxonomic value. Its function is unknown.

cubitus — The cubital vein of the insect wing, vein Cu.

cursorial — Adapted for running: applied as descriptive term for insect legs typically long and slender that characterise many insects that run.

d

decticous — Bearing articulated mandibles; applied to pupae that are able to free themselves from the cocoon by biting.

digging setae — (also: cavisetae) Strongly thickened, sometimes spatulate, setae on female genitalic structures of some Myrmeleontidae, presumed to be used in oviposition.

digitiform — Finger-shaped.

discoidal — Flattened and rounded, disc-shaped.

e

ectoproct — One of paired plates surrounding the anus and derived from tergite x; sometimes fused with tergite ix.

embossed spot — On forewing of some Osmylidae - Spilosmylinae: a small strongly domed area of membrane, usually at posterior margin, and often ornamented by inclusion of elaborated crossveins thickened from their normal state (fig. 50).

empodium — A spiculate or lobelike process (extensionof the unguitractor plate) between the tarsal claws, morphologically distinct from the arolium (q.v.) but often used loosely to refer to the same structure; it can arise from the ventral surface of the arolium.

entoprocessus — (Same term is plural.) In male genitalia: one or both of paired lateral processes from the gonarcus.

exarate — Descriptive term applied to insect pupae in which the appendages (antennae, legs, wings) are free rather than being sealed to the body surface.

f

falcate — Applied to shape of wings, especially the forewing: hooked, sickle-shaped, due to excavation of the margin behind the wing apex.

femur — From the base, the third joint of the insect leg, usually the longest part of the leg, preceding the tibia and tarsus.

filiform — Applied to form of antennae: slender and threadlike, not expanded or elaborated.

forcepate — Appearing like forceps, hence extended and opposable, with implication of grasping. Applied particularly in Neuroptera to expanded ectoprocts or other structures in some males, where used in copulation.

frons — A region on the anterior of the head, delimited posteriorly by the arms of the epicranial suture and anteriorly by the frontoclypeal suture between the antennal sockets.

g *galea* — A lateral distal lobe on the apex of the stipes; part of the maxillae, the second of the three pairs of mouthparts of a typical chewing insect.

gena (genae) — The 'cheek' region of the insect head, between the ventral edge of the compound eyes and the gular suture.

glabrous — Smooth or hairless, used in contrast to 'setose' or 'sculptured' for insect body surfaces.

gonapophysis (gonapophyses) — Components of female genitalia, appearing as ventral or lateral plate-like structures derived from segments viii and ix.

gonapsis — In male genitalia of some Chrysopidae: a predominantly internal structure, with a projecting apex in the dorsal membrane of sternite ix.

gonarcus — In male genitalia: an arch-shaped structure above the aedeagus or other genitalic components.

gonocoxite — A basal sclerite of an appendage modified for oviposition (or, more broadly, of any copulatory structure of either sex).

gonosaccus — An eversible, usually inverted, sac-like membranous structure forming the hind wall of the abdomen and supporting the genitalic structures.

gonosetae — Strong setae, usually in small lateral groups, on the gonosaccus.

gradate vein — One of a series of crossveins. Each series (most commonly across the radial and medial areas of the wing) forming a more or less continuous track across the longitudinal veins.

gula — In the head: a median ventral sclerite, often associated with prognathous head orientation, between the labium and occipital foramen.

h *humeral crossvein* — The basal costal crossvein in the wing, often referred to particularly in the forewing where it can be normal or recurrent (with the apex strongly reflexed or curved basally toward the costal margin).

hyaline — Transparent, colourless (as in wing membrane), as opposed to opaque or coloured.

hypandrium — The subtending subgenital plate below the male genitalia.

hypandrium internum — In male genitalia: a small, often inconspicuous, sclerotised organ at the base of the ejaculatory duct.

hypocauda — In female Berothidae: a long digitate process from the lower margin of each lateral gonapophysis.

hypognathous — Applied to the head orientation in which the head is vertical and the mouthparts directed ventrally.

hypomere — A slender ventral projection from the male phallobase, the basal region of the intromittent organ.

hypostigmatic cell — The cell immediately subtending the pterostigma in Myrmeleontoidea, hence the cell behind the point of fusion of Sc and R1.

i

intramedian cell — In forewings of many Chrysopidae (terminology: im): a small cell near the central wing base, formed by branching of M followed by uniting branches to enclose the cell.

intraradial cell — Any cell within the radial area of a wing, bounded by crossveins between any two branches of R and Rs.

j

jugal lobe — A small lobe at the posterior base of the wing, behind the jugal vein (fig. 66).

jugal vein — A small vein in the jugal lobe of the wing.

l

lacinia — The median sclerite attached to the distal part of the stipes of the maxilla; thus, part of the mouthparts, usually bladelike.

linguella — In female genitalia of some Ascalaphidae: a more or less sclerotised plate between the ventral gonapophysis valves, protruding posteriorly from them and apparently forming the floor of the genital chamber.

m

macrotrichia — Strongly developed setae on the wing membrane or veins, also named trichoid sensilla.

maculae — Spots or other more or less raised circular markings.

media — The fourth of the main longitudinal veins (M) of the insect wing.

mediuncus — In male genitalia: a median posteriorly directed process from the upper part of the gonarcus.

mesonotum — The upper surface of the second segment of the insect thorax.

metanotum — The upper surface of the third segment of the insect thorax.

moniliform — Beadlike. Applied to form of antenna in which flagellar segments are separated by more or less distinct constrictions, rather than being of uniform thickness.

n

nygma (nygmata) — Small spot, presumed to be sensory, on the wing surface. The position of nygmata can be very characteristic in position on the membrane of some Neuroptera and thus useful in identification.

o

ocellar tubercle — Raised triangle on vertex of head, on which ocelli or putative ocelli are sited.

ocellus (ocelli) — The 'simple eye/s' of many adult insects, most often three in number and positioned in the angles of the 'Y' of the epicranial and frontal sutures.

orthognathous — Referring to the head shape in which the primary axis of the head is at a right angle to that of the main body of the animal.

oviruptor — A sclerotised, often toothed, structure on the cephalic membrane of the prehatching larva of some Neuroptera, used to penetrate or cut through the egg chorion at hatching; an 'egg-burster'.

P *papillae (papilla)* — Small, usually digitate or domed, raised prominences.

parabaculum — In male genitalia of Hemerobiidae: term used, following Oswald (1993a) for the structures commonly termed parameres, reflecting that they are not homologous with gonocoxites ix as in other Neuroptera.

parameres — In male genitalia: lateral paired rodlike sclerites, arising close to base of penis. See also parabaculum.

pectinate — Comblike, applied to antennae where the flagellar segments are expanded, or to ornamented tarsal claws.

pelta — In male genitalia of Ascalaphidae and Myrmeleontidae: a small structure with sensory hairs, sited between the apices of the parameres.

pleurostoma — In many Ascalaphidae: a small plate above the base of the mandible.

plicaturae — In adult Coniopterygidae - Aleuropteryginae: small exserted organs present on some abdominal sternites. The number and position can be of taxonomic value in diagnosing genera.

praegenitale — (or *pregenitale*) In female genitalia: a small single or paired structure between the apex of sternite vii and the subgenitale.

presectoral crossvein — A crossvein traversing cell R1 before the origin of the radial sector (fig. 181). The number is sometimes of taxonomic value, as in the hindwing of Myrmeleontidae, where this feature is used in separation of major subfamilies.

prognathous — Referring to an orientation of the head in which the jaws are anterior and the primary head axis is horizontal; commonly found in active predators.

pronotum — The dorsal surface of the first thoracic segment.

prothorax — The first (most anterior) of the three thoracic segments; never with wings.

psammophily — 'Sand-loving'; living in sandy habitats, as do many antlion larvae.

pseudocubitus — In wings of Chrysopidae (terminology: Psc): a vein which appears as the cubitus, but which is formed of branches of Rs, M and CuA.

pseudomedia — In wings of Chrysopidae (terminology: Psm): a vein which appears as the media, but which is formed by fusion of branches of Rs and M.

pseudopenis — In some male Chrysopidae: a single tubelike organ externally on the wall of the gonosaccus.

pterostigma — A pigmented spot/cell/region on the anterior margin of the wing toward the apex, usually at or near the apex of R1.

pterothorax — The 'wing-bearing thorax'; a collective term for the second (mesothorax) and third (metathorax) thoracic segments.

pulvinus (pulvini) — In many male Ascalaphidae: a saclike protrusion from each side of the gonosaccus, usually with long gonosetae. It may protrude considerably from the end of the abdomen when the gonosaccus is everted.

r

radial area — The space in an insect wing between the subcostal and radial veins.

raptorial — Grasping, applied to the forelegs of Mantispidae and few other taxa which are modified for catching and holding prey.

repagula — Rodlike bodies, specialised aborted eggs which accompany the egg batches of many Ascalaphidae.

rugose — 'Wrinkled'; applied to the texture of an insect body surface.

s

sagittal seta — A long seta in the median longitudinal plane, used here to denote conspicuous seta on the clypeus of some Hemerobiidae and useful in classification.

scape — The basal segment or joint of an insect antenna.

scarabaeiform — Resembling the larva of a scarab beetle: an oligopod larva with greatly enlarged abdomen and reduced mobility. In Neuroptera applied descriptively to the larva of some Ithonidae.

seta (setae) — Hollow, bristle-like cuticular projection(s) on the insect body or appendages, often sensory and appearing as small hairs.

setose — Bearing setae.

spatulate — Spoon shaped, narrow at the base and rounded or expanded at the apex.

spermatheca — The receptacle for sperm in the female.

sternite — A sclerotised ventral part of a segment of the insect body; a subdivision of a sternum, and sometimes used loosely as equivalent to the whole sternum (cf. tergite).

styli — Small elongated processes, here usually referred to for processes on the gonapophyses in female genitalia.

subanale — In female genitalia: the sternite xi, covering the anus ventrally.

subarcus — In male genitalia (some Osmylidae): a transverse, arched structure above the parameres.

subgenitale — In female genitalia: the subgenital plate.

sulcus — A furrow or groove in the insect body surface; an externally visible line resulting from infolding of cuticle.

superprocessus — In male genitalia: the dorsal, posteriorly directed process from each paramere.

symplesiomorphy — Sharing of one or more primitive character states by two or more species.

synapomorphy — Sharing of one or more derived character states by two or more species.

synscleritous — Joining or fusion of tergite and sternite of the abdomen to form a complete ring, as in male genitalia of some Coniopterygidae.

t *tentorial pits* — Two pairs of external depressions in the insect head, marking the external signs of the internal head skeleton, the tentorium.

tergite — A sclerotised dorsal part of a segment of the insect body; a subdivision of a tergum (cf. sternite).

tibial spurs — (In Neuroptera) paired spurs at or near the apex of the tibia. Presence, shape and relative lengths on different legs may be of taxonomic value, for example in Myrmeleontidae.

tignum — In male genitalia of Chrysopidae: an arched transverse structure between the gonarcus and the subanale.

trichobothria — Fine sensory hairs from a cupolate base. In Neuroptera it refers especially to the sensilla on the cercal callus of the ectoprocts.

trichoid sensilla — See macrotrichia.

trichosors — In wings of many Neuroptera: short, stout, setose thickenings of the wing margin between tips of the longitudinal veins.

triungulinoid — The freeliving first instar larva of some parasitic insects (here: Mantispidae) which undergoes substantial change in form once a host is reached.

tuberculate — With small rounded, projecting lobes: tubercles.

v *vena triplica* — In Psychopsidae: the three parallel longitudinal veins near the anterior of a wing, constituted by Sc, R1 and Rs, all linking distally (fig. 9).

venation — The pattern of veins in the wing, often diagnostic for higher groups of Neuroptera and useful also in recognising many genera and species.

vermiform — 'Worm-shaped'; applied to larvae that are basically cylindrical in form and lack conspicuous appendages.

vertex — The top of the head.

wax gland — Any gland which secrete a waxy material. In Coniopterygidae: the hypodermal glands on many parts of the body which secrete the characteristic wax or 'meal'.

Index to names of Neuroptera

This index lists all scientific names, and a few English ones, referred to in the text. Of the taxon names mentioned in the checklist (p. 145), only those above the species level are included. The principal treatment of a family starts on the pages underlined; illustrations of a particular taxon are on the pages in **bold**.

A

Acanthaclisinae — 124, 125, 126, **128**, 129, 130, 174
Acanthaclisini — 124, 125
Acanthaclisis — 129, 130, 174
acer Walker (*Myrmeleon*) — 31, **131**, 132
Acheron — **116**, 118, 120, 171
Acmonotini — 115
Acmonotus — 119
Acratoleon (syn. of *Isoscelipteron*) — 135, 137, 175
Acroberotha (syn. of *Isoscelipteron*) — 69, 70
acuminata Banks (*Megalomina*) — 83
Aleuropteryginae — 39, **40**, 41, **45**, 46, **48**, 49, 147
Aleuropteryx — 47
aleyrodiformis (Stephens) (*Semidalis*) — 44, 150
Allotomyia (syn. of *Megalomus*) — 84
amiculus Fitch (*Hemerobius*) (syn. of *Sympherobius*) — 85
amoena Tjeder (*Heteroconis*) — 47, **48**
Anapsectra — 81
Anisochrysa (syn. of *Mallada*) — 105
Ankylopterygini — 101, 163
Ankylopteryx — 92, **93**, 101, 163
Annandalia (syn. of *Psectra*) — 81
Anotiobiella (syn. of *Hemerobius*) — 81
antlions — 9, 10, 32, 122, 124, 130, 132, 134, 137, 138, 143
Apertochrysa — 97, **99**, 100, 103, 166
Apochrysinae — 86, **89**, 90, 163
Archaemyrmeleonida — 123
Archaeomicromus (syn. of *Micromus*) — 84
armatus (*Myiodactylus*) — 109
Ascalaphidae — 6, 9, 12, 13, 17, 18, 20, 30, 31, 32, 33, 34, 111, **112**, 113, 114, **116**, 171, 193, 194, 195
Ascalaphinae — **112**, 113, 114, 115, 118, 119, 121, 171
Ascalaphini — 115
Ascalohybris — **116**, 118, 119, 171
Asperala — 65, 153
audax Van der Weele (*Formicaleo*) (syn. of *Bandidus vafer*) — 138
australiensis Enderlein (*Helicoconis*) (syn. of *Cryptoscenea*) — 47
Austrochrysa — 93, 94, **95**, 99, 100, 103, 166, 168
Austroclimaciella — **61**, 63, 65, 153
Austromantispa — **63**, 64, 65, 154
Austromicromus (syn. of *Micromus*) — 84

B

Baliga — 132
Bandidus — **133**, 134, 135, 138, 175
barnardi New (*Rapisma*) — **37**, 38
beaded lacewings — 9, 68
Belonopterygini — 102, 165
Berotha — 69, 70, 158
Berothella — 74, 159
Berothidae — 6, 9, 10, 12, 13, 14, 17, 19, 30, 31, 33, 36, 68, 69, **71**, 192
Berothinae — 69, **71**, 158
bistrigatus (Rambur) (*Distoleon*) — **136**, 137
Bofia (syn. of *Nuglerus*) — 143, 179
Borbon (syn. of *Dendroleon*) — 141
Boriomyia (syn. of *Megalomus*) — 84
borneensis Meinander (*Paraconis*) — 47
borneole Aspöck & Aspöck (*Isoscelipteron*) — 70
Bornia (syn. of *Borniochrysa*) — 104
Borniochrysa — 34, 93, **98**, 99, 101, 104, 166
brasiliensis Schneider (*Chrysopa*) (syn. of *Plesiochrysa*) — 105
Brauerobius — 81
Brinckochrysa — 98, 100, **104**
Brisus — 137, 175
brown lacewings — 75
Brucheiser — 41
Bullanga (syn. of *Mossega*) — 141, 175
Buxtonia (syn. of *Notobiella*) — 81

C

Calomantispinae — 60
Campanacella — **61**, 65, 154
Campion — 62, 63, 66
canifrons Navás (*Formicaleo*) (syn. of *Bandidus*) — 138
carnea (Stephens) (*Chrysoperla*) — 104, 106
Carobiinae — 84
Carobius — 77, **78**, 84
celebensis MacLachlan (*Myrmeleon*) — 132
Centrolysmus (syn. of *Spilosmylus*) — 56
Chrysocerca — 97, **98**, **101**, 104, 166, 168
Chrysopa — 90, 101, 102, 103, 104, 105, 132, 167
Chrysoperla — 21, 92, **98**, **101**, **104**, 106, 167
Chrysopidae — 8, 9, 12, 13, 15, 19, 20, 21, 30, 31, 32, 33, 34, 56, 85, 86, 87, **88**, **89**, 90, **93**, **95**, **96**, **97**, **98**, **99**, **101**, 102, **104**, 106, 192, 193, 194, 196
Chrysopidia — 94, 104, 105
Chrysopinae — 87, 90, 92, **93**, **95**, **96**, **97**, **98**, **99**, **101**, 102, **104**, 106, 163
Chrysopini — 103, 166
Chrysotropia — 105, 169
Colinus (syn. of *Neurocolinus*) — 134
Conchylosmylus (syn. of *Spilosmylus*) — 56
concinnus Stephens (*Hemerobius*) (syn. of *Wesmaelius*) — 84
Coniocompsa — **45**, 46, 147
Coniocompsini — 46
Coniopterygidae — 6, 10, 12, 13, 14, 17, 20, 21, 29, 30, 31, 32, 33, 34, 39, **40**, 41, **43**, **45**, 47, **48**, 49, 85, 194, 196
Coniopteryginae — 39, **40**, 41, 42, **43**, 44, 49, 149
Coniopterygini — 44
Coniopteryx — 42, **43**, 44, 149
Conwentzia — 39, 42, **43**, 44, 150
Cosina — **128**, 129, 130
Cryptoscenea — **45**, 46, 47, 49, 147
Cymothales — 138, 139, 141, 143, 176
Cyrenoberotha — 69
Cyrenoberothinae — 69

D

Dasypteryx (syn. of *Isoscelipteron*) — 70
decipiens (Roepke) (*Semidalis*) — 44
decisus Walker (*Hemerobius*) (syn. of *Neuronema*) — 84
decoratum Handschin (*Neuroleon*) — 135, **136**
decrepitus Walker (*Ascalaphus*) (syn. of *Idricerus*) — 118
Dendroleon — 138, 139, **140**, 141, **142**, 176
Dendroleoni — 123
Dendroleoninae — 123
Dendroleonini — 123
Dendroleontinae — 124, 125
Dendroleontini — 124, 126, **127**, 134, 138, 139, **140**, 141, 143, 144
diaphanus (Gerstaecker) (*Thaumatosmylus*) — 56
dicax (Walker) (*Helicomitus*) — 119
Dichochrysa — 94, **96**, 97, 100, 103, 105, 169
Dictyochrysa — 90
dietrichiae (Brauer) (*Suhpalacsa*) — 120
difformis McLachlan (*Suhpalacsa*) (syn. of *Suphalomitus*) — 120
Dilar — **73**, 74, 159
Dilaridae — 8, 11, 12, 13, 14, 15, 17, 19, 30, 31, 33, 72, **73**, 74
Dilarinae — 72, **73**, 74
Dimarini — 129
diptera Burmeister (*Hemerobius*) (syn. of *Psectra*) — 81
disjunctus Banks (*Neuroleon*) — 135
dispar Banks (*Acratoleon*) — 137
Distoleon — 134, 135, **136**, 137, 176
Distoleontini — 124, 126, **127**, **133**, 134, 135, **136**, 137
Drepanacra — **78**, 85, 159
Drepanacrella (syn. of *Drepanacra*) — 85
Drepanacrinae — 85
Drepanicinae — 60
Drepanoconis — 47, **48**, 148
Drepanomina (syn. of *Megalomina*) — 83
Drepanopteryginae — 84
duskywings — 39
Dyshemerobius (syn. of *Hemerobius*) — 81

E

Echthromymicini — 129
Echthromyrmex — 125, **127**, 129, 174
Echthromyrmicinae — 129
Eidoleon (syn. of *Distoleon*) — 137
Eidoporisminae — 32, 54
Encycoposini — 115
Eophanes — **133**, 135, 137, 176
Episalus — 34, 137, 139, **140**, **142**, 143, 176

erythraea Brauer (*Mantispa*) (syn. of *Asperala*) — 65
Eucarobius (syn. of *Psectra*) — 81
Euclimacia — **61**, 62, 66, 154
Eumantispa — 65, 66, 155
Eumicromus (syn. of *Micromus*) — 84
eumorphus Yang & Yang (*Navásius*) (syn. of *Dichochrysa*) — 105
evanescens McLachlan (*Nothochrysa*) (syn. of *Evanochrysa*) — 102
Evanochrysa — 94, 102, 165
exigua (Navás) (*Necyla*) — 66
exilis McLachlan (*Idricerus*) (syn. of *Protidricerus*) — 118

F

flavipes Leach (*Ascalaphus*) (syn. of *Suhpalacsa*) — 120
flavum Banks (*Acratoleon*) — 137
Formicaleo — 134, 138
Formicaleon (syn. of *Distoleon*) — 134, 137, 177
formicarium Linnaeus (*Myrmeleon*) — 132
formosa Banks (*Eophanes*) — 137
Frawalkeria — 69, 70, 158
frontalis Burmeister (*Myrmeleon*) — 32, 132
fulvicephalus Scopoli (*Osmylus*) — 58
fulvum Costa (*Isoscelipteron*) — 70
fundatus Walker (*Myrmeleon*) (syn. of *Heoclisis*) — 130
fuscata Fabricius (*Acanthaclisis*) — 130
fuscatus Fabricius (*Hemerobius*) (syn. of *Sisyra*) — 52
fuscinervis Wesmael (*Himanopterus*) — 11

G

Gama (syn. of *Pseudoformicaleo*) — 134
Ganchetus (syn. of *Notobiella*) — 81
gibba Navás (*Nousera*) — 119
Glenochrysa — 56, 96, **97**, 100, 105, 169
Glenurus — 138
Glyptobasis — **116**, 118, 119, 172
gracilis Klug (*Myrmeleon*) (syn. of *Pseudoformicaleo*) — 134
Grammosmylus (syn. of *Spilosmylus*) — 56
grandis Banks (*Dilar*) — 74
green lacewings — <u>86</u>, 87, 92
Grocus — 132

H

Hagenomyia — 32, **131**, 132, 143, 177
hamiltonella Westwood (*Mantispa*) (syn. of *Campanacella*) — 65
handlirschi Navás (*Madrastra*) — 130
Haplogleniinae — **112**, 113, 114, 117, 171
Haploglenius — 114, 172
Helicomitus — 115, **116**, 119, 120, 172
Heliosmylus (syn. of *Spilosmylus*) — 56
Hemerobiidae — 9, 12, 13, 15, 19, 21, 30, 31, 33, <u>75</u>, **76**, 77, **78**, **80**, **82**, 83, 84, 85, 194, 195
Hemerobiinae — 81, 84
Hemerobius — 37, 38, 52, 77, **80**, 81, **82**, 84, 85, 91, 102, 159
Hemerodromia (syn. of *Hemerobius*) — 81
Heoclisis — **128**, 129, 130, 174
Hesperoleon — 137
Heteroconis — 32, 41, **45**, 46, 47, **48**, 49, 148, 149
Heteroleon (syn. of *Bandidus*) — 138
Holoconiopteryx — 42
Holophthalminae — 114
humilis McLachlan (*Drepanepteryx*) — 85
humulinus Linnaeus (*Hemerobius*) — 81
Hybris (syn. of *Acalohybris*) — **112**, 119
Hybrisini — 118

I

Idiomicromus (syn. of *Micromus*) — 84
Idricerus — **116**, 117, 118, 171
ilsae Stitz (*Stenomantispa*) — 66
imbecilla Gerstaecker (*Mantispa*) (syn. of *Austromantispa*) — 65
improbus (Walker) (*Stenares*) — 124
inclusus (Walker) (*Stiphroneura*) — 130
indica (Westwood) (*Mantispilla*) — 52, 66
indica Needham (*Sisyra*) — 52, 66
indicus Navás (*Nepsalus*) — 141
indistinctum Navás (*Neuroleon*) — 135
Indochrysa (syn. of *Semachrysa*) — 102
Indoleon — 135, 137, 177
Indomicromus (syn. of *Micromus*) — 84
infrequens Lambkin (*Toolida*) — 66
insignitus Navás (*Nobilinus*) — 91
insimulans Walker (*Ascalaphus*) (syn. of *Helicomitus*) — 119
insolita Walker (*Berotha*) — 70
insolitus (Walker) (*Nepsalus*) — 141
ismayi New (*Stenogymnocnemia*) — 137

Isoscelipteron — 29, 30, 69, 70, **71**, 158
italicus Rossi (*Hemerobius*) (syn. of *Italochrysa*) — 102
Italochrysa — 87, 94, **95**, **96**, 102, 165
Ithonidae — 8, 9, 10, 11, 12, 13, 15, 18, 32, 33, 38, 195

J
jacobsoni van der Weele (*Chrysocerca*) — 104, 134
jacobsoni van der Weele (*Pseudoformicaleo*) — 104, 134
javanus (Banks) (*Dendroleon*) — 141
javanus Burmeister (*Ascalaphus*) (syn. of *Ascalohybris*) — 119
Joguina — 34, **89**, 91, 163

K
Kempyninae — 32, 54, 58
kervillei Navás (*Nothochrysa*) (syn. of *Stigmachrysa*) — 103
Kimminsia (see also *Wesmaelius*) — 84
Kimminsiella (syn. of *Psectra*) — 81
kiungaensis New (*Dendroleon*) — **140**, 141
Kostka — 34, 94, **95**, 105, 169
Kulinga (syn. of *Neuronema*) — 85

L
laceratus Gerstaecker (*Periclystus*) — 143
lacroixi Fraser (*Indoleon*) — 137
Leptoleon (syn. of *Myrmeleon*) — 132
lestagei Lacroix (*Cymothales*) — 143
Leucosemidalis — 44
leveri (Kimmins) (*Isoscelipteron*) — 70
longus Walker (*Ascalaphus*) (syn. of *Acheron*) — 120
loquax (Walker) (*Acheron trux*) — 120
lorianus Navás (*Nomes*) — 141
louiseae Banks (*Heoclisis*) — 130

M
maclachlani van der Weele (*Acanthaclisis*) (syn. of *Cosina*) — 130
maculata Monserrat (*Sisyra*) — 52
maculata Navás (*Bofia*) (syn. of *Nuglerus*) — 143
Madagascaroleon (syn. of *Dendroleon*) — 141
Madrastra — 129, 130, 174
malayanum Barnard (*Rapisma*) — **37**, 38
Mallada — 94, **96**, **97**, 100, 103, 105, 167, 168, 169
malus Walker (*Myrmeleon*) (syn. of *Stenogymnocnemia*) — 137
manca Gerstaecker (*Mantispa*) — 67
Mansellibertha — 69
Mantispa — 62, **64**, 65, 66, 67, 156
Mantispidae — 6, 8, 9, 12, 13, 14, 18, 30, 31, 33, <u>59</u>, 60, **61**, 62, **63**, **64**, 67, 69, 191, 195, 196
Mantispilla — 65, 66
Mantispinae — 60, 61, 62, **63**, 64, 65, 153
marmoratus Banks (*Dilar*) — 74
marmoratus Banks (*Rexavius*) — **73**
matsumurae (Okamoto) (*Semachrysa*) — 102
Megalomina — 79, 83
Megalominae — 84
Megalomus — 78, 84, 160
Megaloptera — 2, 40
meinanderi Monserrat (*Coniocompsa*) — 46
Menopteryx — 85
Mesohemerobius — 81
Metaconiopteryx — 42
metallica Enderlein (*Parasemidalis*) (syn. of *Neosemidalis*) — 44
Microminae — 83, 84
Micromus — 31, 77, **78**, 79, **82**, 83, 84, 160
Mimetispa — 62, 66, 157
minjerribae Lambkin (*Spaminta*) — 66
minuta Brooks (*Semachrysa*) — 102
mira Navás (*Leucochrysa*) (syn. of *Signochrysa*) — 102
mirabilis Gerstaecker (*Cymothales*) — 143
Mixomicromus (syn. of *Micromus*) — 84
morpheus (Kirby) (*Distoleon*) — 137
Morter — 132
Mossega — 138, 139, 141, **142**, 177
Mucropalpus (syn. of *Hemerobius*) — 81
Myiodactylus — 32, **108**, 109, 110, 170
Myrmeleon — 31, 32, 130, **131**, 132, 134, 137, 138, 141, 143, 177
Myrmeleoni — 123
Myrmeleonidae — 125
Myrmeleontidae — 6, 7, 9, 10, 12, 13, 17, 18, 30, 31, 32, 33, 34, 113, 121, <u>122</u>, **123**, 124, 125, 126, **127**, **128**, 129, **131**, **133**, **136**, 138, **140**, 190, 191, 194, 196
Myrmeleontinae — 124, 125, 126, 175
Myrmeleontini — 124, 126, **127**, **131**, 132
Myrmeleontoidea — 9, 10, 193

N

nacaratus Navás (*Kostka*) — 105
Nallachiinae — 72, **73**, 74
nebulosus McLachlan (*Myiodactylus*) — 109, 110
necrota Banks (*Chrysopa*) (syn. of *Nodochrysa*) — 103
Necyla — 64, 65, 66, 157
Neglurus — 141
Nemoleonini — 123
Nemopteridae — 6, 11, 12, 14, 18, 32, 33
Neomyrmeleonida — 124
Neosemidalis — 42, **43**, 44, 150
Nepsalus — 139, 141, 143, 178
Neurocolinus — 134, 178
Neuroleon — 135, **136**, 137, 178
Neuronema — 78, 84, 161
Neuropterida — 2
Neuropteroidea — 2
Neurorthidae — 11
nevadensis Rambur (*Dilar*) — 74
Nevrorthidae — 8, 10, 11, 12, 17, 19, 33, 34, 52
nicobarica Brauer (*Apochrysa*) (syn. of *Joguina*) — 91
nicobaricum (Navás) (*Isoscelipteron*) — 29, 70
nietneri Hagen (*Dilar*) — 74
nietoi Monserrat (*Conwentzia*) — 44
nigrata Navás (*Chrysopidia*) — 104
Ninga (syn. of *Neuronema*) — 85
Ninguta (syn. of *Neuronema*) — 85
Nobilinus — 34, **89**, 91, 163
Nodochrysa — 96, 103, 166
Noius — 79, 83, 161
Nomes — 138, 141, 179
Norfolius — 110
Nosybinae — 69
notata Navás (*Chrysopa*) (syn. of *Retipenna*) — 102
notatus (Navás) (*Retipenna*) — 102
Nothancyla — 90
Nothochrysinae — 32, 87, **89**, 90, 91
Notiobiella — 77, **78**, 81, **82**
Notiobiellinae — 81
Nousera — **116**, 118, 119, 172
novaeguineensis (Meinander) (*Cryptoscenena*) — 47
nubecula (Gerstaecker) (*Pseudoformicaleo*) — **133**, 134
Nuglerus — 137, 139, 143, 179
Nymphidae — 8, 12, 13, 15, 17, 30, 31, 32, 33, 34, <u>107</u>, **108**, 109
Nyrma — 69
Nyrminae — 69

O

obscurus (Westwood) (*Idricerus*) — 118
oceanicus Navás (*Noius*) — 83
octomaculata (Fabricius) (*Ankylopteryx*) — 101
orientalis McLachlan (*Echthromyrmex*) — 129
Orientispa — **64**, 65, 66, 157
ornata Enderlein (*Heteroconis*) — 47
Osmylidae — 6, 8, 12, 13, 15, 19, 30, 31, 32, 33, 34, <u>53</u>, 54, **55**, 56, **57**, 58, 191, 195
Osmylops — 32, **108**, 109, 110, 170
owl flies — <u>111</u>
Oxybiella — 83

P

pagana Fabricius (*Mantispa*) — 66
pallida Banks (*Zachobiella*) — 81
Palpares — 124, 174
Palparinae — 124, 125, 126, 129, 174
Palparini — 32, 123, 124, **127**, 144
pantherinum Fabricius (*Myrmeleon*) (syn. of *Dendroleon*) — 141
papuensis Meinander (*Coniopteryx*) — 44
Paraconis — 34, **45**, 46, 47, 49, 149
Paraglenurus — 135, 138, 179
Parasemidalis — 44
pardalis (Fabricius) (*Tomatares*) — 124
pardus Rambur (*Palpares*) — 124
partita Enderlein (*Euclimacia*) — 66
peri Tjeder (*Chrysopa*) — 104
Periclystus — 32, 139, **140**, **142**, 143, 179
Phanoleon — 138, 179
phantoma Banks (*Berothella*) — 74
philippinensis van der Weele (*Helicomitus*) (syn. of *Protacheron*) — 120
philippinicola Aspöck & Aspöck (*Isoscelipteron*) — 70
philippinus Navás (*Colinus*) (syn. of *Neurocolinus*) — 134
pictifrons Gerstaecker (*Myrmeleon*) — 132
piepersii van der Weele (*Berotha*) — 70
pineticola Enderlein (*Conwentzia*) — 44
Pirionus (syn. of *Megalomus*) — 84
placidus (Gerstaecker) (*Helicomitus*) — 119
placidus Gerstaecker (*Myiodactylus*) (syn. of *Osmylops*) — 109

Planipennia — 2, 41
platypterus McLachlan (*Echthromyrmex*) — 129
pleasing lacewings — 72
Pleomegalomus (syn. of *Megalomus*) — 84
Plesiochrysa — 31, **93**, 94, **98**, 100, 101, 105, 106, 168, 170
Polystoechotidae — 11, 12, 15
Porisminae — 32, 54
princeps (Gerstaecker) (*Suhpalacsa*) — 120
Proctolyra — 113
Protacheron — 34, **116**, 118, 120, 172
Protidricerus — **116**, 117, 118, 171
Protiobiellinae — 69
Protoconiopteryx — 42
Protoplectrini — 124, 126, **133**, 134
Protosmylinae — 54
Psectra — 77, **78**, 81, **82**, 162
Pseudochrysa (syn. of *Chrysocerca*) — 104
Pseudoformicaleo — **133**, 134, 179
Pseudomallada (syn. of *Dichochrysa*) — 105
Pseudoplectron (syn. of *Pseudoformicaleo*) — 134
Pseudoptynx (syn. of *Nousera*) — 119
Psychopsidae — 11, 12, 15, 18, 32, 33, 196
pulchellus Banks (*Carobius*) — 84
punctata Banks (*Zachobiella*) — 81

Q

quadrituberculata Westwood (*Mantispa*) (syn. of *Austroclimaciella*) — 65

R

ramburi (Schneider) (*Plesiochrysa*) — 31, 106
Raphidioptera — 2
Rapisma — **37**, 38, 147
Rapismatidae — 11, 12, 13, 15, 18, 28, 30, 31, 33, 34, 36, **37**, 38
reticulata Navás (*Mossega*) — 141
Retipenna — **99**, 100, 102, 164
Rexavius — **73**, 74
Rhachiberothidae — 10, 12, 14, 17, 60, 69
Rhachiberothinae — 14, 69
Rhipidosmylus (syn. of *Spilosmylus*) — 56
rubellus Navás (*Campion*) — 66

S

sagax (Walker) (*Hagenomyia*) — 132
sahbanum Barnard (*Rapisma*) — **37**, 38
salomonicus Navás (*Brisus*) — 137
samoana Esben-Petersen (*Austrochrysa*) — 103
scalaris Navás (*Nuglerus*) — 143
Schizophthalminae — 114
Scoliochrysa (syn. of *Austrochrysa*) — 103
scopifer Gerstaecker (*Myrmeleon*) (syn. of *Paraglenurus*) — 138
Scotoconiopteryx — 42
Semachrysa — **99**, 100, 102, 164
Semidalis — 42, **43**, 44, 150
Semohemerobius (syn. of *Hemerobius*) — 81
Sencera — 92, **93**, 101, 164
sexguttata Enderlein (*Spiloconis*) — 47
shirozui Nakahara (*Cercomantispa*) (syn. of *Orientispa*) — 66
Signochrysa — 102, 168
simulatrix McLachlan (*Mantispa*) (syn. of *Mimetispa*) — 66
Sinoneuronema (syn. of *Neuronema*) — 85
Sisyra — **51**, 52, 151
Sisyridae — 8, 9, 12, 13, 17, 30, 31, 33, 50, **51**, 52, 190
Sisyrina — 51, 52
somnolentus (Gerstaecker) (*Distoleon*) — **136**, 137
Spaminta — 65, 66
spectrum Rambur (*Palpares*) — 124
Spiloconis — 46, 47, 49, 149
Spilosmylinae — 32, 54, **55**, **57**, 58, 151, 191
Spilosmylus — 30, 54, **55**, 56, **57**, 151
Spinomegalomus (syn. of *Megalomus*) — 84
spinosus New (*Umbranymphes*) — 110
spongeflies — 9, 50
Stenares — 124, 174
Stenogymnocnemia — 135, 137
Stenomantispa — 63, 66, 157
Stenomicromus (syn. of *Micromus*) — 84
Stenosmylinae — 32, 54
stigma Girard (*Hemerobius*) (syn. of *Synthochrysa*) — 91
Stigmachrysa — 94, 103, 166
Stigmatosmylus (syn. of *Spilosmylus*) — 56
stigmatus Navás (*Mallada*) — 105
Stilbopteryginae — 32
Stiphroneura — 130, 174
strenua Gerstaecker (*Mantispa*) (syn. of *Tuberonotha*) — 67
Suhpalacsa — **112**, 115, **116**, 119, 120, 172
Suhpalacsini — 115, 118
Suhpalasca (syn. of *Suhpalacsa*) — 120

Suphalasca (syn. of *Suhpalacsa*) — 120
Suphalomitus — 115, **116**, 119, 120, 173
suzukii Okamoto (*Mantispa*) (syn. of *Eumantispa*) — 66
Sympherobiinae — 85
Sympherobius — 77, 80, 85
Symphrasinae — 60
Synthochrysa — 34, **89**, 91, 163

T
tacitus (Walker) (*Indoleon*) — 137
tacitus Walker (*Myrmeleon*) (syn. of *Indoleon*) — 137
Tahulus (syn. of *Pseudoformicaleo*) — 134
Tanca (syn. of *Micromus*) — 84
Thaumatosmylus — 34, 54, 56, **57**, 153
timidus Hagen (*Micromus*) — 31, 84
tineiformis Curtis (*Coniopteryx*) — 44
Tmesibasini — 114
Tomatares — 124, 125, 175
Toolida — 62, **64**, 65, 66
tortricoides Rambur (*Megalomus*) — 84
traceyae New (*Coniocompsa*) — 47
Triadochrysa (syn. of *Mallada*) — 105
Trichomatinae — 69
Triplochrysa — **89**, 90
tristis Hagen (*Myrmeleon*) (syn. of *Hagenomyia*) — 132
trux (Walker) (*Acheron*) — 120
tuberculatus Walker (*Osmylus*) (syn. of *Spilosmylus*) — 56
Tuberonotha — **61**, **63**, 67, 157
typica Esben-Petersen (*Glenochrysa*) — 105

U
Umbranymphes — 34, **108**, 109, 110, 170
umbrosa Navás (*Chrysopa*) (syn. of *Apertochrysa*) — 103
uniformis Navás (*Neuroleon*) — 135

V
vafer (Walker) (*Bandidus*) — 138
Vaja (syn. of *Notobiella*) — 81
valdezi Banks (*Chrysopa*) — 102
variegatus Fabricius (*Hemerobius*) (syn. of *Micromus*) — 84
venusta Hagen (*Chrysopa*) (syn. of *Ankylopteryx*) — 101
verticalis Banks (*Distoleon*) — 137
vesiculigera Enderlein (*Coniocompsa*) — 46
vigana Navás (*Sisyra*) — 52
viridipennis Walker (*Hemerobius*) — 37, 38
viridis Tillyard (*Notobiella*) — 81
vitripennis (Navás) (*Dendroleon*) — 141

W
Wesmaelius — **80**, 81, 84, 162
westermanni Esben-Petersen (*Protacheron*) — 120
winkleri Navás (*Bornia*) — 104

X
Xaviera — 64, 67, 158
Xeroconiopteryx — 42, 44, 149

Z
Zachobiella — 77, **78**, 81, **82**, 83, 162
zephyrinus Gerstaecker (*Episalus*) — **140**, 143

F A

M A L E